Technology of Food Product Development

First Edition

Other books edited or written by the same author –

Chemistry and Technology of Cereals as Food and Feed
First and Second Editions

Glossary of Milling and Baking Terms

Bakery Technology and Engineering
First, Second, and Third Editions

Equipment for Bakers

Cereal Technology

Cereal Science

Food Texture

Water in Foods

Bakery Technology

Ingredients for Bakers

Cookie and Cracker Technology
First, Second, and Third Editions

Technology of the Materials of Baking

Snack Food Technology
First, Second and Third Editions; Japanese Edition

* * * * * *

TECHNOLOGY OF FOOD PRODUCT DEVELOPMENT

by

SAMUEL A. MATZ, PH. D.

President, Pan-Tech International, Inc. Formerly, Vice President for Research, Development, and Compliance, Ovaltine Products, Inc. At one time, Vice President for Research and Development, Robert A. Johnston Co.; Technical Director of the Refrigerated Dough Program, Borden Foods Co.; Chief of the Cereal and General Products Branch, Quartermaster Food and Container Institute for the Armed Forces; Chief Chemist, Harvest Queen Mill and Elevator Co.; Instructor, Department of Flour and Feed Milling Industries, Kansas State University; Chemist, Iglehart Mills Division of General Foods.

PAN-TECH INTERNATIONAL, INC.
P. O. BOX 4548
MC ALLEN, TEXAS 78502
1994

Phone & Fax 956-702-7085

Technology of Food Product Development

ISBN 0-942849-29-9

Pan-Tech International, Inc.
P. O. Box 4548
McAllen, TX 78502

Printed in the United States of America

PREFACE

The development of new food products constitutes a large part of the history of the advance of civilization. Who can doubt that the simple discovery of boiling dry seeds in water led to the development of grain storage and permitted the coalescing of wandering tribes into towns and villages where other, more complex, improvements in life could take root and flourish. Then, as porridges of the most primitive type developed into beer and leavened breads, the steady march of cereal technology began, to reach at last the perfection of Bud Lite and chocolate-coated granola.

New food products can contribute many benefits to the life of the average consumer. There will always be those who complain about Twinkies and the like, but they fail to consider the many moments of innocent and harmless pleasure that such inexpensive foods contribute to the life of the average consumer. For those who prefer foods engineered to contribute fully balanced nutrition in each mouthful, there are ample opportunities to satisfy their needs. And, for those persons obsessed with dietary fads, there will always be a continuous stream of more pleasurable alternatives to sun-dried fruits and rancid nuts. The important factor is, individuals in this country can still apply their own values to food selection – in most cases. There is, however, a continual effort by those who "know what is best for us" to force standards on the general population – standards often based on faulty, incomplete, or biased knowledge.

In a free society, the right to choose or reject foods based on individual standards of flavor, appearance, cost, nutrition, and packaging (in other words, on the bases of personal taste and affordability) is as much a fundamental right of each person as is the right to free speech, free exercise of religion, and a free press. And, it is a right that can be exercised more frequently and with greater personal enjoyment than many others. Those who attempt to place rigid and/or illogical guidelines on the availability or production of new foods, show a contempt for personal choice and individual responsibility at least as great as those who would deny food service to someone on the basis of his or her ethnic affiliation.

I believe the development of new food products is of advantage to both industry and consumers. In the great majority of food manufacturing companies with which I am acquainted – and these range from individual entrepreneurs to conglomerates selling millions of dollars worth of new foods every day – product development is in the hands of responsible individuals. They vary in technical skill and creativity, of course, and sometimes they may be influenced by financial exigiencies, but I am sure they never disregard that important factor, the good of the customers.

Samuel A. Matz
January 24, 1994

Contents

chapter one

THE PROS AND CONS OF NEW PRODUCT DEVELOPMENT

INTRODUCTION

Is it necessary for a food manufacturing company to introduce new products if it wants to expand its sales and profits, or are there better ways of achieving growth? There is no simple, unequivocal answer. Even if we could all agree that an ongoing product development program is generally productive of growth and enhanced profitability, our conclusions could not be expected to apply to all types of food producers; different companies have different goals and varying capabilities. For example, some firms may be unable to institute product development programs even if they wanted to do so simply because their money and manpower resources cannot support any activity that does not result in immediate production, sales, and profits. These companies are terminal cases, all of their resources being required merely to stay out of bankruptcy and nothing is left for new product programs or any other type of investment in future growth. For them, new product development is but a dream.

Other companies may decide that their expertise in marketing and distribution is so great that only token efforts in making innovations are needed in order for them to continue to function in a profit-making mode. They are often mistaken.

A company may base its expansion policy chiefly on reverse engineering of products successfully introduced by competitors, achieving sales by pricing the imitations substantially cheaper than the products being sold by the competitor. Problems involve possible infringement of patents, trade secrets, or trade dress, and the necessity for cutting profits to the bone in order to achieve the lower prices required by this kind of merchandising.

In spite of the negatives, there is widespread agreement that new product development is suitable, and perhaps essential, for progressive, well managed companies willing and able to devote time and money over a period of many months or even years to a continuous, well organized effort aimed at expanding markets and improving profitability. This view is shared by most food manufacturers, as is shown by the data in Table 1.1, (based on Anon. 1992A) which shows an increasing rate of new products entering the marketplace. It is reasonably certain that the companies making these introductions would not have committed so much time, effort, and funds to new product development unless their experience and their analysis of future prospects had shown them that new product development was essential to their survival and growth.

Table 1.1

RECENT TREND OF INTRODUCTIONS OF NEW FOOD PRODUCTS

Year	Number	Change from previous year
		%
1984	5,396	--
1985	5,617	4
1986	6,107	9
1987	7,866	29
1988	8,183	4
1989	9,192	12
1990	10,301	12
1991	11,846	15

DOES YOUR COMPANY NEED NEW PRODUCTS?

When you begin to consider the role you want new products to play in shaping your own company's future, it is a good idea to first answer the question, "Does our firm need to introduce new products in order to expand its sales and profits?" If the owners are doing well with their current line of products, would it not be better to spend their available funds on improving production efficiency and expanding distribution rather than run the risks and experience the problems of developing and introducing new products? If the company's distribution of existing items is not nationwide or worldwide, would it not be better to use the company's capital to expand its markets geographically rather than try to capture new kinds of business?

It is probably true that many companies selling staples or commodity-type goods (sugar, salt, flour, etc.) could get along fairly well for many years without introducing any new items, in spite of all the advantages generally claimed for product development. This is particularly true if the company owns well-known and respected tradenames. Such companies are, however, risking gradual erosion of their market share as their competitors probe for ways to displace their command of consumer dollars, and a technical or marketing breakthrough by a competitor can cause a sudden collapse of sales of the old-line company. Real growth of such firms – except for the possibility of geographical expansion – is usually very difficult, slow, and costly in the absence of the introduction of successful new products. An additional fact to to consider is that the profit margin of staple items is almost always very low and usually is in a long-term downward trend, while new products often yield relatively high margins.

Several authors have commented that products go through sales (or consumption) cycles that are in some ways analogous to the life cycles of organisms. If this hypothesis is correct, the stages in the life of a product can be described, for example, as testing, introduction, market expansion, market saturation or stability, and obsolescence or death. Arguments for this sequence of events have not been universally accepted, but, if the analogy does hold true, a decline in sales is inevitable for all products, and constant introduction of new products seems to be necessary if a company is to continue in business for more than a few years.

Small companies often regard product development as being too costly for their kind of operation. Actually, it appears that the most economical way for many small companies to significantly increase sales is by introducing new products. This is due to the nature of the competition – although large companies commonly can spend massive amounts of money on advertising and promotion, can achieve greater economies in purchasing, production, and distribution, and can hire better sales personnel than small companies, they do not necessarily have better idea evaluation and consumer testing capabilities, and these are crucial factors in new product development.

A small- or medium-size company is certainly at a disadvantage relative to the large multi-plant operators insofar as the size of its technical workforce concerned, but this inferiority need not extend to the effectiveness of its R&D group since its efforts can be concentrated on a few important products instead of being spread over a wide spectrum of projects. The ability to move fast when an opportunity is recognized, by substituting intuitive management decisions for test market evaluations, can often greatly reduce the amount of laboratory activity required to develop a marketable product.

When we look at the history of the food industry in the U.S., it appears reasonably clear that new products are very important factors promoting long-term survival and growth. We all know of companies whose fortunes have declined and others that have gone out of business because they failed to develop new products or introduced poorly conceived and inadequately tested items. On the other hand, we can point to companies whose meteoric rise is due entirely to the introduction of products that met a previously unfilled consumer desire. These are extreme examples; it is more common to see trends that are more erratic and less clearly defined.

The food industry has benefited from a substantial amount of creative product development activity during the last several years. Most of this work was conducted by large diversified food companies with central development staffs, but worthwhile advances were also made by small, specialized manufacturers who found niches of consumer demand not satisfactorily filled by competitors. Some innovative new products have been based on work done initially by the technical service personnel of suppliers of ingredients, equipment, or packaging materials and passed on to a food manufacturer willing to give some quid pro quo in the nature of future ingredient purchases, etc.

If a company is going to start (or continue) a product development program, top management must insist upon detailed advanced planning and rigorous control of the projects. The principles involved in efficiently managing new product development are reviewed in subsequent chapters.

WHAT TO EXPECT FROM AN NPD PROGRAM

To evaluate the effect new products could have on your company's prospects, present and expected future conditions should be analyzed. The following steps, slightly modified from an article by Rives (1964), can be helpful in making this evaluation:

1. Prepare a long-range (5 to 10 years) forecast of sales of existing product lines.
2. Prepare a long-range profit plan for the company based on the sales projections derived in step 1.
3. Prepare an inventory of the company's capabilities and resources, including but not restricted to (a) means of distribution, (b) geographical limitations, (c) financial limitations, (d) technical staffing limitations, (e) legal limitations (such as antitrust requirements) and (f) existing processing lines (and particularly any incompletely utilized processing capabilities).
4. Determine the most favorable market opportunities for new products based on the company's capabilities and resources.
5. Prepare a modified long-range profit plan, incorporating the probable effect of new products.
6. Prepare a statement of the company's new objectives, including an estimate of the extent to which sales and profits will be expanded by the introduction of new products.
7. Assign responsibility for new product development and formulate procedures for establishing projects.
8. Provide for evaluation of new product performance.

This generalized scheme will not be entirely satisfactory for most companies, but it can serve as a basis for designing plans better adapted to your specific requirements.

LEARNING FROM THE PAST

A great many articles dealing with the cost, success rate, and influence on company profits of new product development programs have been published. It is difficult to correlate or even interpret most of these publications because of uncertain terminology (there is not much agreement even on the meaning of "new product," for example) and bias (what size or type of companies are being considered, what type of products, etc.). Furthermore, one can seldom feel confident about the validity of the data, if any, cited in the articles. Time is also a factor, the quantitative criteria of success in, for example 1970, would probably not be the same as for a new product introduced in 1994. In the present section, "Learning from the Past," the author will attempt to extract from the literature some reliable guidance from the more useful articles and to apply his own experience to the subject.

Definition of New Products

A very significant problem in understanding media reporting of new product occurrences, trends, and forecasts, is deciphering the meaning to the author of "new products." It has to be said that the definition varies even among separate works from the same author and often in various parts of the same publication. One publisher that conducts an annual survey says they use the following definition: "A new product is one that the consumer has never been able to purchase in the U.S. before this particular item was introduced. We eliminated new flavors, sizes, packages, cooking alternatives, product improvements, new ingredients, new processing, etc." There are ambiguities even in this relatively detailed definition. Some other usages of the term are explored in the following paragraphs.

(1) Old foods in new packages are very frequently reported in company records and the media as "new products." New sizes, such as two pounds of product in each package instead of (or in addition to) the old one-pound pack, would be called, by some, a "new product." Adding a baking pan to the paperboard carton of an existing baking mix package would almost always be considered a new product. Improving the performance of a bag for microwave-poppable popcorn by adding a novel kind of susceptor would probably result in a new product. The creation of a different item number in the company's ordering system would probably signify a new product to Marketing, and certainly would to Distribution – packaging the consumer units 18 to a case rather than 12 in a case would therefore be regarded as the introduction of a new product.

Changes in label graphics can sometimes be regarded as generating a new item. Perhaps a newly-appointed head of the Marketing Department, after overseeing the designing of a new logo – a common exercise for such executives – would call all packages to which that logo is affixed, "new products."

(2) Food products that are changed in dimensions often may reasonably be described as new products – very small or very large cookies may differ in acceptability and usage characteristics from the conventional sizes – but is a bread loaf sliced 16 to the pound a new product as compared to one sliced 24 to the pound? Is a carbonated beverage in a 3-liter bottle a different product even though a 2-liter bottle of the same formula beverage has been sold for years? In such cases, a valid decision would have to be based in part on the consumer's view of the change.

(3) Many items designated as "new products" are merely line extensions of existing foods; this leads to new products having very minor changes in composition from preceding versions, such as new flavors of gelatin desserts, or blue-coated M&Ms instead of red-coated ones, or canned tuna without salt, or a hot cocoa mix with marshmallows instead of without, or Oreo cookies with more filling than the previous model. It is true that some of these modifications require a great deal of technological and engineering work, and, from that standpoint might justifiably be called "new products." One has to consider whether the criteria for applying this nomenclature will be based mostly on technological considerations or at least partly on consumer perceptions.

(4) It is very common to call an old food offered for new uses a "new product." Frozen pancakes in a microwavable container would be new, even though frozen pancakes of the same size and formulation in an ordinary freezer carton have been marketed for years. If technological changes are involved, no doubt the product would be entitled to the description even under the cleanest definition.

(5) A product that has been out of the market for several years, when reintroduced (often with some minor changes) is almost always described as a new product by the Marketing and Sales Departments.

On the other hand, many companies will hesitate to call a significantly improved food sold under an existing label and trademark a new product if the improvement does not noticeably affect appearance, performance, or taste, even though the production cost has been reduced, the production process has been improved or simplified, better ingredients are being procured, or storage life has been improved.

Success Rates in New Product Development

There have been many studies purporting to quantitate the number or percentage of new products that achieve success in the marketplace. The conlusions reported in most of these studies are of doubtful value to managers planning a development program because of the variable criteria that reporers used for determining the success or failure of products, the difficulty in obtaining reliable data from the companies that were surveyed, the wide variability of success ratios in different industries, and many other factors. However, the commitment of small companies to continuing programs for new product development has probably been significantly inhibited by published statements that 90% (a commonly quoted figure of uncertain provenance) of all new products fail.

Most of the very high failure rates quoted in published studies appear to use as a base figure 100% of all the product ideas that have been given any kind of consideration by a company. If the base figure is instead restricted to projects yielding prototypes suitable for some kind of consumer test, the failure percentage drops dramatically. Hopkins (1980) suggested that the failure rate for new products actually introduced into the marketplace is much lower than the earlier mentioned 90% figure, and is probably nearer 33%. His survey included many different kinds of products, not just retail food items, and it is recognized by the author that the failure rate of the latter type of product could be significantly higher than the overall average.

Other criteria can be used to determine "success" and will affect the prospects apparently faced by a company starting a new product development program. In one survey (Abrams 1980), a product was defined as successful it it met "management's original expectations for it in all important respects."

The calculated success rate is going to be very small if the success ratio is based on the percentage of the number of new ideas officially entered into a planning system that result in products paying back their investment costs and returning a net profit within, say, five years. If, on the other hand, one bases the standard of success on technically perfected products that achieve some level of stable market penetration within a reasonable time after adequate expenditures for marketing have been made, the success ratio should be much higher. Examples of some of the most recent studies will be discussed briefly.

According to a survey made by a company called Group EFO (Anon. 1992B), an estimated 86% of the 15,401 new products introduced in 1991 failed to reach their business objectives. In 1990, the figure was 90%. The study found that 81% of the new products were merely extensions of existing brands. Introductions of new foods dropped 11% from 1990 to 1991.

A publication by Drennan (1987), described the results of a survey, part of which are shown in Table 1.2.

Table 1.2

PERCENTAGE OF PROJECTS REACHING COMMERCIAL APPLICATION

Respondents	Success rate
%	%
14	<5
22	6-10
7	11-20
14	21-30
7	31-40
16	41-50
20	50+

Other estimates (Abrams 1980) include: 98% failures, a figure attributed to Dancer Fitzgerald Sample (minimum yearly retail sales of $15 million regarded as successful); about 67% failures according to A. C. Nielsen; and 33% failures as estimated by The Conference Board. Booz, Allen & Hamilton, Inc. found that only one in 58 new product ideas becomes commercially successful (Bengston 1982). These latter figures include a wide variety of new products, not only introductions by food companies.

James Behnke, senior vice president of technology for Pillsbury (reported by Best 1991) stated he expected a 60% fallout from the brainstorming stage to test market and, of the 40% that survive, about 70% will succeed. This gives about 28% successful new products based on what appears to be a relatively unevaluated idea pool as 100%. This is considerably larger than most published estimates.

At the opposite extreme is a statement by Sloan (1993) that in 1992 83% of products brought to market were projected to miss their business objectives. Allowing for some difficulty in determining the meaning of "business objective," a 17% success rate for products actually introduced into the market place seems unduly low. Not surprisingly, she concludes, "New products were not a high priority in most corporations, seeking growth via investments in established brands, followed by downsizing to control costs, were the preferred strategies."

Cost of Industrial R&D

If companies are presumed to follow some rational plan in allocating their funds, the behavior of experienced management in financing new product development should be an indication of the importance of such operations. One of the best and most extensive reviews of R&D investment plans and experience is that of Prepared Foods magazine (O'Donnell 1993). In this article it was reported that 10,319 new food products had been introduced in 1993 through September, as compared to 9,891 and 9,415 in the same periods of 1992 and 1991, respectively.

The difficulty of separating the cost of basic (or semi-basic) research from the cost of developing individual products has been remarked by many students and is a real problem confronting financial analysts and company management. It is probably true that genuine research of a type that could be called "basic" is rather rare among food companies: most CEOs would feel quite uncomfortable about funding technical activities that are not connected even remotely to a visualizable new product or product line. For small companies and, even more, for individual entrepreneurs, "research" of a basic nature cannot be contemplated.

Sperber (1993) published a report on R&D expenditures of food companies (actual and projected) as a percentage of sales dollars. This material is summarized in Table 1.3. The picture is clouded, however, by the statement that most (about 70% of respondents) included quality control and quality assurance costs in the R&D budget.

The data in the Sperber table indicate that, in the panel consulted (which appeared to be biased toward larger companies), around 80% of the respondents expected to spend 1% or less of their sales dollars on R&D. For some of the companies, only part of this expenditure would be applied to new product development. If this pattern is carried through to smaller companies, it follows that firms having less than five million dollars in sales could not afford even one technologist devoted to new product development.

Table 1.3

R&D BUDGET AS A PERCENTAGE OF FOOD SALES

R&D Budget, % of sales	Percent of companies reporting		
	1991	1992	1993
Under 0.5	51	43	48
0.5 to 1.0	33	38	30
1.0 to 1.5	9	13	11
1.5 to 2.0	7	3	4
2.0 to 2.5	0	3	7

OPPORTUNITIES FOR INDIVIDUALS AND SMALL FIRMS

Most persons having an entrepreneurial disposition take an optimistic attitude toward new product development, but they are likely to become discouraged when they learn about the costs to large companies of these programs. It is certainly true that new product development can be expensive and time-consuming, and the payoff may never arrive. The president of Pepperidge Farms said, in 1979, that it took his firm three to five years with an investment of three to five million dollars to get any new food product into a good profit position. An average lag, for all industry, of 1.2 to 2.5 years between research expenditure and flow of of initial revenues, was estimated. It is difficult to say how the intervening years have affected the time and cost parameters in this particular company, but they can hardly have decreased. It is not too far-fetched to assume that costs and time for other companies would fall in the same ranges.

CEOs of small companies in the food industry realize that such costs and such long lead times for idea materialization far exceed their resources. They must operate under guidelines that are much different from those of big companies, and must minimize or eliminate all unnecessary steps; many of their decisions must be made on the basis of informed intuition, rather than on reams of data accumulated by scientists, lawyers, and marketing people. Risks must be taken, but the willingness and ability to take risks constitutes a large part of the strength of the individual entrepreneur.

Personal experiences and observations tell us that small operators can triumph in the marketplace, even in head-to-head competition with the largest food companies. Though the most publicized and most striking new product introductions seem to be those of the large mass marketers, there have been many less well-known successes attributable to the efforts of individual entrepreneurs. At first glance, the deck certainly seems to be stacked in favor of large operators, but there are some factors favoring individuals (and small companies) who have good ideas, determination, a willingness to learn and adapt, enthusiasm for work, and (of course) some capital. Luck is needed, too.

The incredible maze of federal, state, and municipal regulations affecting food products is a serious hindrance to the small operator. The hidden agenda behind the enactment of this tangle of restrictions is that such a hindrance is intended by the lobbyists, their employers (large corporations and industrial associations), and their lackeys (legislators and bureaucrats), to restrict competition. The expenses of designing products to meet all pertinent regulations, making analyses to insure legal composition of individual products, labels, and plants, liasing with regulatory bodies, and fighting completely unworkable regulations in court are simply passed on to the retail buyer as part of the price by the manufacturers, distributors, and supermarket chains that dominate the market. On the other hand, the small company with its low sales volume will often find it impossible to finance the initial costs that are incurred, and when the product's retail price is marked up to cover cost of compliance it cannot be afforded by the average consumer. In this way, the market becomes effectively closed to small operators who might be able to provide new products consumers want.

Yet, in some respects the earliest phases in the introduction of a new item may be easier for the small independent operator than for the multiplant food manufacturer. The latter will have to spend perhaps hundreds of thousands of dollars for analyses, evaluations, label design, legal advice, etc., merely to insure that one product meets the requirements of all jurisdictions in which it will be introduced, and most of this money must be laid out before the product is introduced anywhere. The individual entrepreneur or small company that aims to exploit first a limited market may find the task somewhat simpler. Some of these innovators will, in practice, skip a few of the theoretically essential compliance steps. If, however, the product receives a good initial response and it is decided to expand the market beyond an inconspicuous local distribution area, serious (even disastrous) conflicts with regulatory agencies may then be encountered.

Timing is another factor in which the smaller firm may have some advantages. It has often been said that agreements of department heads are more rapidly arrived at in small companies, and that misunderstandings between parties responsible for taking action are more quickly resolved. It has been my experience that this is not always the case. But, if the CEO of the small company is decisive and forceful, it may very well be true that red tape and wheelspinning can be kept to a minimum, so that product development progresses more rapidly to a definite conclusion.

The success of a sole proprietorship, partnership, or small corporation just starting out in the food industry will depend to a great extent on the instincts of its CEO as applied to consumer acceptance and marketing strategy. Extensive consumer testing of flavor, appearance, and texture will not be economically possible for these companies, and product potential will often have to be evaluated on the spur of the moment by the entrepreneur. Charles Lubin of Sara Lee, Jeno Paulucci of Chun King, Jeno's pizza, and other enterprises, and Harlan Sanders of Kentucky Fried Chicken fashioned at least their initial product lines according to their own tastes and/or their perception of consumer tastes. In spite of early mistakes and failures, they became phenomenally successful, but there have been thousands of failures by CEOs who used the same criteria and found, too late, that their tastes were out of step with the preferences of most consumers.

OBJECTIVES AND PLAN OF THIS BOOK

The book you are reading has as its main objective the enlightenment and guidance of scientists, technologists, engineers, and allied technical people working in all phases of new product development. The mission is, keep them from making serious mistakes and improve the efficiency of their activities. In addition, there is a special concentration on methods of new product development available to individual innovators and managers of small companies.

There are fifteen chapters of which this, the first chapter, is presented as an introduction to the subject of new product development and as a justification of that area of work. At the end of this chapter is a lengthy bibliography, extending into 1993, that lists many articles the reader may wish to consult to further his or her understanding of the reasons for and against new product development. This bibliographic plan is followed in each of the chapters and represents one of the most useful features of the book.

The remaining fourteen chapters deal with theory, principles, and examples of new product development with heavy concentration on technical matters of concern to the food industry.

Chapters two and three deal with methods and plans for originating new product concepts; how to get ideas and how to evaluate them.

Chapters four and five concern projects, how to set them up and how to follow and control their progression from idea to finished product.

Chapter six contains discussions of various aspects of the actual performance of the development tests, sample production, and data evaluation during the laboratory and pilot plant work leading up to the processing of market test samples.

An important part of any successful project is the efficient and diplomatic handling of contacts with other company departments and with outside agencies and suppliers. Some suggestions on interfacing are covered in chapter seven.

Chapter eight is reserved for discussions of problems concerned with containers, especially the design of packages and the materials available to the designer.

In chapter nine, the designing, manufacturing, and applying of labels is discussed. This leads into chapter ten, which covers the very difficult and often confusing area of regulations affecting label graphics and text. A discussion of the new labeling regulations issued by the FDA and USDA early in 1993 and of industry response to those regulations is included.

Chapter eleven is devoted to topics related to market testing, including the economical and timely preparation of samples, interpretation of sales tests, and modification of products to meet consumer requirements.

Many of the operations in product development can be simplified and improved by the judicious use of computers. The pros and cons of computer use at various stages from the generation of ideas to the interpretation of sales test data and the instrumentation of production lines is treated in chapter twelve.

If the new product is significantly different from those being offered by competitive firms, management will be interested in preventing other firms from copying it. Chapter thirteen describes several ways of gaining legal protection for innovations made by your company.

Chapter fourteen analyzes the present direction of new product development in the food industry and projects these lines of innovation to the near and distant future. Readers may gain some inspiration for specific new products from these rather general predictions.

In chapter fifteen, you will find thirteen "commandments" (brief cautionary statements) summarizing the many pitfalls surrounding the complex and variegated practice of new product development.

BIBLIOGRAPHY

ABRAMS, B. 1980. Study of new-product failures refutes basic premise of growth. Wall Street J. (June 26), p. 23.

ANON. 1968. How much should your company spend on R&D. Business Management *34*, No. 2, 57-60, 62.

ANON. 1984. Product optimization modeling, common sense can solve wide range of research problems. Marketing News *17*, No. 3, 6.

ANON. 1990. KGF emphasizes fast action in new product development. *19*, No. 19, 1-4.

ANON. 1992A. New product success: How are the definitions shifting? Gorman's New Prod. News Spec. Ed., Jan. 1992, p. 4-7.

ANON. 1992B. New-product troubles have firms cutting back. Wall St. J., Jan. 13, 1992, page B-1.

ANON. 1993. Smart research planning pays off. R&D Magazine *35*, No. 11, 13.

BELSHAW, F. 1978. General Mills exec explores the mission of R&D. Food Product. Dev. *12*, No. 5, 52, 54, 56, 58, 60.

BENGSTON, R. A. 1982. Nine new product strategies: Each requires different resources, talent, research methods. Marketing News *15*, No. 19, 7.

BENJAMIN, B. 1993. In the ice cream R&D trenches. *54*, No. 9, 95, 97.

BEST, D. 1989. Industry and government seek R&D synergy. Prepared Foods *158*, No. 2, 84-86, 88.

BEST, D. 1991. Pillsbury "greens" an R&D giant. Prepared Foods *160*, No. 3, 53-56.

BROWN, J. K. 1984. The Impact of Technological Change on Corporate and R&D Management. The Conference Board, NYC

BUSS, D. D. 1993. Nabisco's Warren: New products that take the cake. Food Processing *54*, No. 10, 67-68.

CARLSON, K. O. 1977. New products – problems or opportunities? Food Prod. Dev. *11*, No. 7, 34, 36, 38.

DESROSIER, N. W. and DESROSIER, J. N. 1971. Economics of New Food Product Development. AVI Publishing Co., Westport, CT

DRENNAN, W. C. 1987. R&D: Entering an era of challenge. Food Eng. *59*, No. 5, 91-93, 95-97, 101-102.

FRAND, E. A. 1978. How to build a lightning rod – part 1. Industrial Res. & Dev. *20*, No. 3, 17.

FRAND, E. A. 1982. Getting into and out of trouble. Industrial Res. & Dev. *24*, No. 7, 29.

FRIEDMAN, M. 1989. Hard questions, soft answers. Prepared Foods *158*, No. 8, 23, 25, 27.

GIRAGOSIAN, N. H. 1980. Successful Product and Business Development. Marcel Dekker, NYC

HIMMELFARB, P. A. 1993. New product development. Today's Chemist at Work *2*, No. 4, 56-57.

HOPKINS, D. 1980. Survey finds 67% of new products succeed. Marketing News *13*, No. 16, 1.

HOWARD, N. 1980. Business probes the creative spark. Dun's Review *115*, No. 1, 32-36, 38.

LARSEN, C. K. 1979. The new products illusion. Manufacturing Confectioner *59*, No. 4, 46-48.

LAURO, G. 1977. Product development today. Food Technology *31*, No. 7, 73-74.

LEPKOWSKI, W. 1993. U.S. missing chance to hone competitive edge. Chem. & Eng. News *71*, No. 11, 4.

MARTING, E. 1964. New Products – New Profits. American Management Assoc., NYC

MARVIN, P. 1972. Product Planning Simplified. American Management Assoc., NYC

O'DONNELL, C. 1993. 1994 R&D investment survey. Prepared Foods *162*, No. 12, 38-50.

OTTO, A. 1993. Half-baked ideas that spell success. Food Business *6*, No. 20, 23.

PEDI, R., and MOESTA, R. 1993. Total quality in product development. Prepared Foods *162*, No. 2, 44-46.

RIVES, R. C. 1964. What is your new product policy? *In* New Products – New Profits, E. Marting (Editor). American Management Association, NYC

ROCK, P. A. 1993. Basic, applied research. Chem. & Eng. News *71*, No. 13, 3.

SLOAN, A. E. 1993. Product flops: what every executive should know. Food Business *6*, No. 4, 43.

SPERBER, B. 1993. Foods of tomorrow. Food Processing *54*, No. 8, 36-37, 39-40, 42.

STANLEY, A. O. and WHITE, K. K. 1965. Organizing the R&D Function. American Management Association, NYC

SULLIVAN, D. 1976. Good product concepts embody business goals, market needs. Food Product Dev. *10*, No. 4, 28, 30.

SWIENTEK, B. 1993. Hershey "Hugs" team building. Prepared Foods *162*, No. 11, 44-45.

TIJUNELIS, D., and CLAUSEN, N. 1979. R&D on a Minimum Budget. Anacom, NYC

WALTERS, J. 1993. Building the best new products in America. Food Processing *54*, No. 11, 16-17, 20, 23, 27-28, 30, 32, 35-36, 38, 40, 42.

WALTERS, J. E. 1965. Research Management: Principles and Practice. Spartan Books, NYC

WHITE, V. P. 1988. Handbook of Research Laboratory Management. ISI Press, Philadelphia, PA

chapter two

GENERATING A STOCK OF IDEAS

INTRODUCTION

The generation of ideas is a normal and more or less continuous function of the human mind, but generating ideas that will ultimately result in successful new products is a considerably rarer and more difficult activity. It is to be expected that psychologists and other persons of that ilk would study ways of facilitating idea production. Among the rather large number of publications that have described the habits, patterns, environment, or attitudes most conducive to the generation of productive ideas is found the following list (Anon. 1989):

(1) Seek variety.
(2) Challenge the rules.
(3) Never take the first answer as the only solution.
(4) Write down ideas as fast as they come to you.
(5) Use analogies.
(6) Try things. Make mistakes and use them to lead you to other ideas.
(7) Reverse the situation – visualize a solution that does the exact opposite of what is required.
(8) Listen to music. Soft music . . . helps you relax so you can be more creative.
(9) Learn from others. Build on others' ideas that partially solve the problem.
(10) Verbalize ideas.

Although most of us would disagree with one or more of the items in the preceding list, it at least serves as a basis for making up our own set of rules for facilitating the production of ideas that can ultimately be turned into new foods having good consumer acceptance.

The next best thing to originating a good idea of your own is to have someone else do it for you, preferably at a nominal cost. The literature describing ways to facilitate the generation and/or accumulation of new product ideas by employees or their accumulation by other means is extensive. Many of these articles are, however, so elementary that they have little practical value. In other cases, though the author's conclusions may seem theoretically sound, they are difficult or impossible of application in a corporate environment. In a few instances, authors have taken into account the realities of industrial R&D and have suggested apparently feasible methods for acquiring a stock of ideas to which can be applied a selection or evaluating process. A number of these articles are reviewed in the following discussion. Cited publications are not restricted to the food field.

SOURCES OF IDEAS

There is a widespread belief in the food industry, and perhaps in other industries as well, that the collecting of the largest possible number of ideas is a highly desirable preliminary to the initiating of a research project. Since any technologist and all departments have limitations on the amount of time that can be spent in evaluating ideas, it is obvious that some restrictions must be placed on the stock of ideas that must be evaluated later. The smaller the organization, the more selective it must be in order to correlate the time available for idea gathering with the requirements of the company for new products.

It has frequently been advocated that suggestion forms be distributed to all employees, in the hope that a startling new idea will be elicited. If such forms are used, there should be a requirement that the submitter give reasons why he believes the product or process will be worthwhile. Nearly all suggestions obtained in such surveys, will, in the author's experience, be for products that have already been considered by management.

In a survey of over 150 American companies, Booz, Allen, and Hamilton found that 88% of new product ideas came from within the companies. Marketing and R&D personnel contributed 60%. There are many other sources for ideas. Several authors (Hilton 1961, Marvin 1972, Marting 1964, Ammerman 1963) have published extensive lists. One report (Kill 1965) suggests: (1) Internal sources: sales, advertising, and merchandising staffs; R&D personnel; production and engineering departments; and employee suggestions. (2) External sources: consumers' suggestions; trade suggestions from wholesalers and retailers; suppliers' suggestions; competitive products, both domestic and foreign; technical and trade publications; and patents.

Ammerman (1963) listed the following possible sources of new product ideas: (1) company suggestion plans, (2) requesting ideas from the sales staff and providing a place on sales reports for entering these ideas, (3) contests both inside and outside the company, (4) U.S. government publications such as the "Product List Circular" published by the Small Business Administration, (5) product scouts employed to look for new product ideas both abroad and in the U. S., (6) industrial research firms, (7) market research firm reports, (8) advertising agencies, (9) journals and other publications, (10) consultation with suppliers, (11) trade fairs, and (12) plant tours.

To stimulate the flow of ideas from both inside and outside the company, Marvin (1972) suggested that very clear and pointed requests for help must be made. Among the possible sources of new ideas listed by this author are:

1. Venture capital corporations interested in financing the development of new products.
2. Directors of research in other corporations that may make new product ideas available on a royalty basis.
3. Newspapers, trade journals, and magazines describing new products.

4. Organizations and individuals offering to search patent records for ideas that can be developed.
5. Professional societies and business organizations having staff members specializing in certain segments of business activity.
6. Purchasing agents knowing of special market requirements.
7. Inventors with potentially marketable ideas.
8. Public utility companies interested in developing commercial opportunities for their services.
9. Manufacturers' representatives familiar with new developments.
10. Government services related to scientific, industrial, and military activities.
11. City, county, and state development commissions charged with industrial growth of specific areas.
12. Patent attorneys representing client inventors of new products.
13. Trade associations interested in specific areas of industrial activity.
14. Advertising agencies creating merchandising programs for products to be farmed out for manufacturing.
15. Commercial bankers directing the financing of new product activities.
16. Consulting engineers working on projects in the idea and design stage and looking for a commercial tie-in.
17. Sales representatives alert to market needs and potential new developments.
18. Scientists, engineers, and technicians possessing ideas suitable for development.
19. Business acquaintances and personal contacts who are particularly observant of market needs and opportunities for new industrial ventures.

There are other even more extensive lists of sources of ideas. Peter Hilton's Handbook of New Product Development (1961) lists about 100 sources, many of them being relatively minor subdivisions of those discussed previously. A publication of the U. S. Chamber of Commerce (Anon. 1955) includes a checklist of 40 major sources of new product ideas.

"Reminders that might be used as guidelines" were given by Ellington (reported by Marting 1964):

1. Establish new products goals. It is surprising how easily ideas flow when organizations zero in on a target.
2. Organize the search for sources. Do not depend on chance. A network must be established and nurtured lovingly.
3. Keep alert to "what's new" for each industry, material, and geographical location.
4. Use many sources on an imaginative, open-minded basis. Don't limit yourself to a single source.
5. But, when a good source of products is found, keep at it. There are very few highly creative "corners" within people's minds and in business organizations, and once found they can be responsible for most of your company's good ideas.

All alternatives are not equally productive. A list of searches that turned out to be unsuccessful has been given by Marting (1964):

1. A consumer goods manufacturer required every salesman to come up with one idea a day, space being provided on their call sheets for entering these ideas. After reviewing the suggestions for a year, management concluded that the ideas were not well thought out, made little business sense, and — for the most part — were not new.

2. One new product manager, tired of being stopped in the hall by company personnel who had been stimulated by a new ideas campaign, wrote the following in his newsletter. "It doesn't do much good for you to say, 'I think we should get into the missile business. This is going to be a great market in the future. Men will be flying to the moon and back — we should have a part in such a booming business.' This doesn't help much! If, however, a suggestion is made that we should be missile men because: We have the best equipped machine shop in northeastern America; and it is operating below capacity; and it can machine fine parts to close tolerances; and there is a certain location advantage by being in New England; and X million dollars of orders were given last year by firms we can identify — this information would be helpful and meaningful to us. That's an idea." This new product manager got across the message that an idea not compatible with a firm's existing or potential capabilities is not a help but a hindrance. It wastes someone's valuable time.

All of the aforementioned writers are long on description and opinions but rather short on data. As a result, their suggestions can best be regarded as only the most general type of guideline, needing expansion, selection, and fleshing out with details suitable for actual situations.

Competitors may be unwitting sources of ideas. Management sometimes decides it is necessary to reproduce an item already in the marketplace in order to meet competition. The usual idea generation and selection procedures will then be overridden. The originator of the project will often be a sales representative who reports that a competitive item is steadily encroaching on the market share of an existing product of his company.

When a scheme or plan is established for generating ideas, there must be allowances made for overriding it in special circumstances. For example, a strikingly novel and apt idea may suddenly be brought forth by someone within the company and management immediately realizes that it represents a substantial opportunity. In other cases, R&D will realize that a new development in technology makes possible the production of an article devoid of some of the objectionable features of existing counterparts. Each of these new product recommendations must be fully coordinated with all other affected departments before a research project is initiated, and they must compete for priority with new product ideas from all other sources.

Recycling Old Ideas

A review of company files will sometimes unearth product ideas that have been considered previously, perhaps even developed nearly to completion, only to have been dropped because of some negative factor. The circumstances may have since changed so that the negative factor no longer exists, or is of much less importance. For example, the market may have become more receptive to the type of product, or a necessary ingredient was too costly but is now much cheaper, or laws that prohibited or greatly restricted sale of the product may have been revised, or more efficient production techniques may have become available. Many more scenarios leading to an improved outlook for old ideas can be imagined, including the simplest one of all, that the company has become desperate for new products whereas it previously took an ultraconservative view of product introductions.

Reviving old projects may be somewhat complicated by the existence of personal prejudices and enthusiasms within the management group. It may happen that one of the department heads is adamantly opposed to the product – perhaps the production vice president thinks it will be too difficult to manufacture or the marketing vice president believes it will introduce unwanted complications in his activities. If the hatchet man who put down the idea at the time of its original introduction is still around, the idea will likely be blasted again, but if he has since departed there may be a chance to complete the development work and add an item to the company's product line with relatively modest expenditures of time and money.

On the other hand, it is very common to find someone on the team who has a fixation on some product, perhaps because it is similar to a food remembered from his childhood, and who persists against all evidence in urging reincarnation of the dead project.

Although some circumstances may have changed for the better since the old project became dormant, it is also possible that even more serious problems have developed in the interim. It is very unwise to accept all of the preceding work as still being valid without extensively reviewing each aspect of the project. For example, the proposed product may have at one time received legal clearance for formula and label claims, but now may be difficult or impossible to market because of changes in laws and regulations.

Reviewing the Literature

Cookbooks. It is desirable for the food technologist who is assigned responsibility for new product development to have access to a large collection of cookbooks. Most municipal libraries have fairly large collections of cookbooks, though there has often been no system guiding their acquisition. Some university libraries have collections which include very old volumes, as do the Library of Congress in Washington and the New York City Public

Library. The technologist should, over a period of time, acquire a personal collection of cookbooks relating to his or her professional interests – baked products, beverages, etc.

The purpose of reviewing cookbooks is not to extract recipes which can be used as is, but to spark interest in approaches hitherto ignored. If there is some possibility of patenting a concept, it is valuable to review cookbooks for prior art. It is often surprising (and disappointing) to discover that a new product idea which seems brilliantly innovative has been described long ago by some cookbook author, but it is far less costly to learn of this prior art before a patent application has been filed or, worse, after a patent has been granted and a competitor has commenced a suit to invalidate the patent.

There are four major kinds of cookbooks being issued in the U. S. today. The first is the encyclopedic cookbook, which attempts to cover the whole range of products the home cook might want to prepare. Often, the recipes in these books will be short, even telegraphic in style, and they will have been obtained from a number of sources, often without verification of their practicality. It is recommended that the technologist have two or three of these cookbooks in a personal library, since they make it easy to look up quickly the basic preparation techniques for any item which may come to mind. It is best to choose books with extensive indexes.

The second type is the "theme" cookbook specializing in a particular kind of product, say on bakery products or Irish cookery or dietetic foods. They usually contain a running commentary, often intended to be humorous or insightful, on subjects of very little interest to the technically trained reader. They may be composed by simply combining recipes taken from other publications (often without prior testing) of the type pertinent to the book's theme. These books are seldom particularly helpful, although an occasional example will show an individual approach and contain some creative modifications. Generally, however, the technologist will be able to make a better investment elsewhere.

The third type is the celebrity cookbook. If the celebrity is a famous chef, the cookbook may be of considerable value and probably will be fairly accurate in the details of ingredient specifications, mesurements, and methods. If the celebrity is a figure from the entertainment or political world, the recipes are likely to be a re-hash of standard recipes and are often very inaccurate and impractical.

The fourth category includes the cookbooks produced for fund-raising purposes by members of a club, church, synagogue, ethnic organization, lodge, or other group. Hundreds of such titles are produced every year. It will generally be found that most, if not all, of the recipes they contain have been clipped from a recent edition of one of the encyclopedic cookbooks even though they are put forward as having been originated by, e.g., "dear Aunt Margaret who died in 1964 at the age of 92 after making these biscuits every day since she was a girl."

Fund-raising cookbooks that concentrate on foods of a specific ethnic group (Italian, Estonian, African, Jewish, etc.) frequently contain interesting recipes that may be adaptable to the tastes of a more eclectic audience. It would not be profitable, or even possible, for the technologist to review all of the available fund-raisers, but a quick glance through those from some of the smaller nationality groups might yield a nugget or two.

Patents. New patents in the technologist's field should be reviewed routinely, using the Patent Gazette as the initial source, then ordering only those patents that seem to have some connection with current needs. The Patent Gazette is too expensive to be included in the budgets of most small laboratories. If there is a Federal Depository Library near you (call your local library for location), a complete run of Patent Gazettes should be available there for your review. If your computer has a modem, and you are connected to **Dialog**, you can obtain access to the "Claims" file, covering U.S. patents back to 1950, and can view the full text of patents granted since 1971 in "U.S. Patents Fulltext." There is also the JAPIO file with three million unexamined Japanese patents, Derwent World Patents Index provides six million international patent abstracts, and INPADOC covers applications and registrations from 56 countries for the latest legal status information.

Ideas for variant and hitherto unpatented approaches to the goals of a disclosed invention may occur to the reviewer of recently issued patents. Also, the invention may show that some product or process on which the reviewer has been working is no longer feasible because of a conflict with the patent. Collateral ideas may be stimulated by reading the prior art discussion in the first part of the patent, even though these discussions are often incredibly confused and inaccurate (perhaps intentionally).

Journals. Of course, scientific and technological journals in food and allied fields should be reviewed soon after they become available. A scientific advance described in a publication of this type, even though it is not patentable or not obviously of practical value, may be have some application to a problem or idea which the reviewer has been considering for some time but has relegated to inactive status because of an apparently insoluble technical difficulty. The discovery described in the journal article may eliminate the need for a great deal of laboratory work and lessen the uncertainty involved in re-activating a project.

In literature searches, semi-technical and business journals should not be neglected. Periodicals such as "Food Processing," "Prepared Foods," "Baking and Snack," "Snack Food," and even "Food Business" frequently contain articles with considerable technical content and the discussions are usually well researched and reliable. These magazines are seldom abstracted, or are inadequately reviewed, for databases and other technical bibliographic resources.

Conventions and Exhibitions

There are many food shows and other commercial meetings at which information on state-of-the-art culinary practices can be obtained, personal contacts made, and products viewed. Shows for the benefit of restaurateurs are held in various parts of the country, but the National Restaurant Show, usually held in Chicago in May, is certainly one of the biggest. Fancy Food Shows, scheduled once a year in New York City and on the west coast are often good sources of ideas, especially if gourmet foods and beverages are the main fields of interest. Less well-known conclaves, such as the Oxford Symposium on Food and Cookery (Levy 1993) may be helpful if time and travel funds are available.

R&D workers will usually find it cost effective to attend the national meeting and exhibit of the Institute of Food Technologists, though not necessarily every year. It is worthwhile mostly for the exhibits of new ingredients. The technical presentations are helpful if you are troubled with insomnia. The American Association of Cereal Chemists and the American Society of Bakery Engineers have small exhibit areas at their annual meetings. Food exhibits in Asia and Europe are sometimes very large in extent and scope, and are often valuable sources of uncommon ideas.

Idea Brokers

Product development consultants are individuals or small groups of people, usually with marketing and advertising backgrounds, who make a business out of peddling ideas or semi-developed products to management. Their sales technique often involves dramatic presentations accompanied by highly optimistic sales projections. Objective evaluations of their worth are hard to come by, but one paper that seems to take a fairly disinterested view is "What Do Users and Prospective Users Think of External Product Development Services" by Mattson and Lott (1971). This article summarizes responses gained from an anonymous survey of seventy food industry product development specialists and from interviews with several consultants engaged solely in new product activities.

Generally, the industrial food product development specialists were unenthusiastic about using consultants to develop new products. When asked to tell the interviewer why they reacted negatively, the respondents gave the following reasons (listed in order of decreasing frequency):

1. Internal staff more capable, and can do as good or better a job than the consultant.

*2. Poor experience with consultants.

3. Duplication of efforts.

*4. Lack of knowledge of specific markets.

*5. Too expensive, costly.

*6. Consultants provide little more than ideas, and this is not a sufficient basis on which to proceed.

7. Lack of confidentiality.

*8. Consultant's work is difficult to internalize.

*9. Consultants lack depth.

*10. Consultants lack facilities.

*11. Clients lose control.

An important feature of the above list is that a number of the reasons (those preceded by asterisks) seem to indicate actual unfavorable experience with consultants, not merely uninformed prejudices.

Consultants, on the other hand, claim (Arenson 1972) that "ideas supplied by consultants are the extraction of related projects with other clients, development specialists, and contemporaries." Potential clients might be tempted to ask if some breach of confidentiality (by the client's standards) is not implied by these "extractions."

Some instances of client dissatisfaction are implied by Arenson (see previous citation) when he says that "no small portion of the reported difficulty in obtaining adequate results from the external consultant can be traced to poor administration and communication failure between client and consultant." Similar problems no doubt lead to failure of the client to obtain adequate results from his internal staff, and the solution might be to improve administrative policies and internal communications rather than to seek outside help.

The main disadvantage of using consultants is that they must sell their specific new product concepts in order to survive, regardless of the idea's compatibility with the client's business. Consequently, they may be tempted to conceal defects, exaggerate advantages, and rely too strongly on the appeal of novelty. "Research," says Barstow Bates (Anon. 1967), president of New Product Services, Inc., "is the process of going up alleys to see if they are blind." There must be a more efficient way to proceed than going up innumerable blind alleys. The R&D scientist might contend that his first duty is to read the street signs in an attempt to avoid a few of the blind alleys.

Copying the Competition

Unfortunately, the new product programs of many smaller companies (and of a few large ones, too) seem to consist entirely of trying to copy successful product introductions made by their competitors. The assumptions on which this approach is based are that the original introducer has done all the consumer research and lab testing which is necessary and these costly parts of the new product development process need not be performed by the copier. The situation is not quite as simple as it appears, but there have been studies indicating that a certain amount of success can be achieved by taking a competitor's novel concept and improving it in some relatively minor way. Carlson (1977) reported that, in a survey of 20 highly successful new products, only four were the first of their type on the market. The other 16 ranged from second to tenth (or even later) in their appearance on retailers' shelves. He concluded that it is better to be second or third in the stores with a first class product than to be first with a second class product. This makes sense if the difference or improvement is recognizable by the consumer as being of significant benefit to him.

Copying is at best a risky proposition, however. The market can be saturated by a new product before your copy of it is ready for sale, patents may prevent duplication, and the operating know-how essential for successful manufacturing may not be achievable without very extensive and costly engineering studies. If a company is to grow, or even to maintain its status quo, in this rapidly changing industry, it is necessary to have an effective product development program that does not rely entirely on copying — even if innovating has to be limited to the part-time efforts of one person. Managers who imagine that their companies can continue to survive by manufacturing commodity-type items to be sold at a lower price than their competitors simply do not realize the contribution technical advancements can make to lower costs and improved sales.

Whether or not one wishes to copy someone else's new products, it is a good idea to keep abreast of competitors' innovations. The common term for this type of operation is "industrial intelligence" and it is by no means restricted entirely to information on technical matters. In fact, marketing activities (advertising, promotion, introductions, etc.) of competitors usually generate more interest at the CEO level than do inventions, new processes, and other scientific and technological developments.

The possible ways of doing industrial intelligence can be divided roughly into two categories, legal and illegal. The dividing line seems to vary among different jurisdictions, but it is generally accepted that it is permissible to study the public actions and utterances of the competitor. Examples of legal and ethical sources of information are speeches, press releases, patents, SEC filings, and court documents emanating from the competitor's executives and other representatives. Public government documents of many kinds will often provide an astonishing amount of detail including much material that would ordinarily be considered sensitive, but weeding the wheat from the chaff and deciphering the information once it has been located is a very time-consuming and often frustrating task.

Information of this kind can also be obtained from clipping bureaus, personal reviews of newspapers and other media, computer searches of databases relating to the field of interest, and many other sources. A review of sources of information on other companies' products and an example of how these sources were used to develop a profile of a specific company's activities can be found in "Company Information: A Model Investigation" (Law-Yone 1980).

Nearly everyone considers it to be perfectly ethical and legally acceptable to "reverse engineer" a formula or process based on an examination and analysis of a product being marketed by the competitor, always with the understanding that processes or products covered by patents can't be copied during the term of the patent. Ingredient statements on labels often provide a basis for reconstructing a formula; the nutritional claims may also be of some help. These data can be supplemented with simple chemical, microscopical, and physical analyses of the food. Microscopic examination often gives some clues to the method of processing, low-power magnification usually being of more help than higher powers.

More questionable, but probably legal, ways of obtaining useful information, are surveillance of the competitor's receiving area (from **outside** their grounds) to determine the type and source of raw materials being used, friendly discussions with executives of competing firms to obtain details on product development that they are willing to release (perhaps with quid pro quo), fraternizing with technical and marketing personnel of the company, using the same advertising agency (often a prolific source of information), seeking out suppliers of equipment and ingredients to the company and pumping them for information, and hiring away some of their personnel (but first, determine what kind of secrecy agreement the employee has signed with his present employer).

Unacceptable from an ethical and probably from a legal point of view are such things as pumping a present employee of the competitor for information under the guise of a pre-employment interview, placing a spy on the competitor's payroll (perhaps in a menial position such as janitor, which gives the spy access to many restricted areas), tapping their phones, filing baseless legal actions in attempts to get information from depositions made by the competitor's employees, using a cooperative journalist to interview company executives on topics of interest to you, and employing dissatisfied employees of the company as part-time consultants.

A survey of company executives (number and qualifications not clearly defined) by Moskal (1977) provided the following information about what the executives considered "ethical" (a very elastic term) ways of learning about competitors. Fifty percent or more of the panelists thought the following methods of gathering intelligence were ethical:

(1) Reviewing published materials and public documents

(2) Camouflaged questioning and "drawing out" of competitors' employees at technical meetings.

(3) Financial reports and brokers' research studies.

(4) Acquisition and analysis of competitors' products.

(5) Legitimate prospective employment interviews with people who worked for a competitor.

(6) Disclosures made by competitors' employees, if obtained without subterfuge.

(7) Market surveys and consultants' reports.

(8) Trade fairs, exhibits and competitors' brochures.

(9) Reports of your own salesmen and purchasing agents.

In the Moskal survey, less than 50% of the panelists considered the following methods of gaining information to be ethical:

(1) Direct observation under secret conditions.

(2) Job interview with a competitor's employee when there is no real interest in hiring him.

(3) False negotiations with a competitor for a license to produce a product.

(4) Hiring an employee away from a competitor to get specific know-how.

(5) Trespassing on competitors' property.

(6) Bribing competitors' suppliers or employees.

(7) Planting an agent on the competitor's payroll.

(8) Eavesdropping on competitors by bugging their phone or office.

(9) Hiring a professional investigator to get a specific piece of information.

The concept of "trade secrets" should be considered briefly in relation to industrial intelligence. A trade secret is information (e.g., a description of a new process or a list of customers) which has been developed within, or acquired by, a company and which the company uses due diligence to preserve as a secret. A trade secret cannot be any information which is publicly announced or which the company does not use reasonable care to restrict to company employees and agents who must know it in order to adequately perform their jobs. Misappropriation of trade secrets by competitors is generally regarded as a crime and the violator can be sued for damages by the injured party.

IDEA GENERATION AS A PLANNED PROCESS

A few commentators have made the point that merely collecting and reviewing ideas originating spontaneously in the minds of a diverse collection of individuals does not make full use of the creative potential available to management. Or, as stated by Kotler (1967), "Companies with a major interest in product innovation or diversification cannot . . . rely on informal and spontaneous processes of idea generation. An increasing number are recognizing the advantages of seeking out new areas of risk systematically." This is easier said than done. Some writers have attempted to give specific methods or procedures that an individual or group can use to generate a multiplicity of ideas on a given subject. Most of these schemes have the merit of at least focusing the action at the point of need and, to some extent, avoiding the wasted effort that results from the uncritical processing of every idea that might appear from any source. Formation of committees composed of supposedly creative persons selected from inside or outside the company has been one of the more common approaches.

Panels Assigned to Generate Ideas

In furtherance of the panel method of idea generation, a number of systems for collecting ideas have been published. For example, Marvin

(1972), dean of professional development at the University of Cincinnati, recommended the use of a new product panel composed of a handpicked group charged with the specific responsibility of developing and collecting new product ideas. Demonstrated creative ability and a diversity of interests would be important qualifications for members of this panel. Five attributes have been found to complement each other in the makeup of individuals suited to serve on a new product panel, Dr. Marvin says. These are: demonstrated creativity, the ability to associate ideas and facts in new relationships, a questioning approach, acute powers of observation, and the ability to concentrate. The goal of the group should be finding rather than screening ideas – of course, even choosing where to look for, or from whom to solicit, new ideas is in itself a kind of screening process.

Members of the panel would be specifically charged with the responsibility for generating new product concepts. Choosing the right people for the panel will insure optimum productivity. It follows that, if such a panel is not very productive, you have not chosen the right people.

Crockett (1969) wrote about the generation of new product ideas specifically for the food industry. He claimed that so-called creative types are not the only sources for good ideas. Many techniques should be used, and the problem areas should be approached from different viewpoints to identify ideas or possible ideas.

Matrix Systems

Tauber (1971) proposed a matrix system based on the assumption that combinations of words can be used to represent product ideas. He claimed that most psychologists accept the definition of creative output (the idea) as combinations of two or more concepts in the mind of the creator. If one could obtain all relevant concepts that apply to the food product areas, then by definition the set of all combinations of these concepts (the power set) would represent the total number of all food product ideas, given today's knowledge.

In conducting tests on his theory, Tauber utilized the following general procedure:

1. The names of as many existing food products as could be identified were broken down (verbally) into components that could be found in many different products.
2. The list of words resulting from the first step was then classified into the various "dimensions" of the food world, relating to the questions:
   a. What? Product form, technology, packaging, ingredients, etc.
   b. How? Consumer preparation, kitchen appliances.
   c. When? The meals at which the food is used.
   d. Why? The consumer benefits which result.
3. Since the combinations made possible by the operations performed in the second step would have exceeded, in their example, two to the five

hundredth power (an enormous number not susceptible to manipulation by normal methods) sets of alternatives most likely to lead to acceptable solutions were searched out by applying a procedure based on the hypothetical principle that the value of any idea was the sum of the value of each factor-word plus any interaction effect.

By assuming that the "heart' of most food product ideas can be described by a two-word combination, search time is reduced considerably, and a number of possibilities are generated to which further dimensions can be added sequentially. A second assumption was that certain cross-classifications result in more interesting ideas than others, e.g., combinations of fruits and vegetables (peas and pears) are less likely (intuitively) to be rewarding in new product concepts than are, e.g., different technologies applied to different food forms.

The most interesting cross-classifications resulting from the preceding steps are generated in matrix form and the participant reviews the grid, crossing out combinations representing an existing product and circling combinations he thinks might have market potential. As might be expected, variable results are obtained by different people because of their differences in knowledge, perception, taste, and creativity. Because the two-word descriptions are incomplete ideas, the operator or participant must use his imagination in adding details, and this is a form of creativity.

Brainstorming

One of the more curious schemes for generating ideas was "brainstorming," a method that was characterized by the uncritical generation of loosely connected ideas, usually by small groups of people operating over several hours in an isolated environment. The operative word is "uncritical." Quantity was the aim. No evaluation of the ideas was supposed to take place in the session. This is the obvious defect in the method, since guidance or selectivity is the sine qua non of problem solving. According to Bartow (1975), brainstorming involves a creative problem-solving session having the steps of (1) developing a focus, (2) developing a creative environment, (3) generating ideas, (4) evaluating and analyzing those ideas, and (5) demonstrating the potential of selected ideas. The last two steps are not actually part of brainstorming sessions, as they are usually conducted.

One group that conducted brainstorming sessions planned on having 200 or 300 ideas at the end of each meeting, and found that as many as a thousand were sometimes generated (Anon. 1967). The person quoted in the referenced article said, "There have to be some good ones in that number [why?]. We've never struck out yet. Our trouble is getting the ideas down to a reasonable number for the client to consider." Of course.

For several years, brainstorming was a very popular technique in some marketing circles and was heavily promoted by several advertising agencies. As stated by Detwiler (1976), ". . . it was, more than any other [fad], the ultimate fulfillment of the iron law of Madison Avenue fads. That the amount of noise, hoopla, and publicity which the fad generates varies directly with its lack of utility." Mr. Detwiler should know whereof he speaks, for he was vice president of public relations at BBDO from 1954 to 1975 and one of his functions in that position was to publicize brainstorming. He says, ". . . it may have been one of my most creative efforts ever, to keep the awful truth about the thing from coming out."

Other authors have taken a more neutral approach than Detweiler did to the evaluation of brainstorming, but they have generally reached about the same conclusions (Bouchard 1969; Taylor *et al.* 1958; Moore and Anderson 1954). Their experimental findings indicate, for example, that individuals working alone are more effective than groups in generating more and better solutions.

Brainstorming was never much in fashion in scientific circles because the deficiencies of the method become apparent very quickly when factual concepts are being dealt with. It did have a tremendous impact for a few years in general management circles and especially at advertising agencies. It now seems to be pretty much dead as an idea generating tool, although, like the "Hula Hoop," occasional revivals can be expected and, in fact, a few favorable mentions of brainstorming have been noticed recently in marketing journals and in business publications.

Focus Groups

Focus groups consist of from 2 to perhaps 12 selected panelists who, led by a moderator, engage in a discussion on a specific product or other topic. Participants are selected according to a set of criteria the market research supervisor considers to be important — sex, income level, family situation, and age are commonly included in the profile used for qualifying the subject. Often, the discussion takes place in a room having a one-way mirror so that clients or other observers can monitor the proceedings without disturbing the panelists.

The participants are usually not told — at least initially — the object of the discussion, which is in most cases the elicitation of or evaluation of ideas for new products. Panelists are generally allowed considerable freedom of expression, but are guided and prompted by the moderator if they stray too far from the targeted subject. The proceedings are often taped (video or audio) for future analysis at the convenience of the moderator or his or her assistants. Samples may be presented to the panelists for viewing or tasting.

It is said that focus groups were based to some extent on psychological group therapy. Marlow (1987) quotes Dietz as saying that "unknown to themselves, people in all walks of life and all occupations tell more about a topic, and in greater depth, if encouraged to act spontaneously instead of reacting to questions." The reader will no doubt see some connection to the rationale for brainstorming. Marlow continues, "Focus groups capitalize on the value of group dynamics by encouraging participants to react to one another's ideas and thoughts. In addition, the moderator can challenge and probe for the most truthful responses, yielding a more in-depth accounting of consumers' thoughts and opinions."

Focus groups may have some value in generating new product ideas or in screening projects which have already been partly developed. The panels should never be regarded as being representative of the general population or even of the targeted consumer group, and, therefore, extrapolation of their results to marketing performance is not justifiable. Comments of the participants are nearly always affected by the presence of the moderator and by the comments and demeanour of their fellow panelists, more so in later than in earlier parts of the discussion, so caution must be used when analyzing the tapes.

It is the author's opinion that focus group "research" has its greatest value in providing seeds of ideas for further investigation. After observing many focus group sessions, he has concluded that no quantitative value can possibly be assigned to their results. An additional possible value of focus groups is in pointing out some serious deficiency in a prototype; enthusiasm of management for a particular concept may have led them to ignore or downplay a feature which, when pointed out by the focus group, becomes an obvious negative feature of considerable magnitude. That is, the focus group gives a fresh view of the prototype, not biased in the same direction the management group is biased.

Creating a Product to Fit an Established Need

The collapse of a comptetitor, or the withdrawal of certain products from the market by a competitor, may create an opportunity to meet a small but steady demand for products no longer available to customers who have developed a liking for these items. Such market vacuums occur when, for example, a large manufacturer has re-assessed the demand for a product it has been supplying; perhaps the reception for a new product or line of products has not met some pre-set goal. If your company has lower standards for success and is willing to be a supplier on a smaller scale, it may be possible to introduce a similar or improved product that meets the existing demand. Maybe there are economic, cultural, or geographical niches the original company does not think it is worthwhile to fill.

Among these opportunities are the vacancies created when a merged company that has been selling a product, or lines of products, with a loyal but limited following, decides it is no longer economically feasible to service this small and/or thinly concentrated market, and abandons the tradename as well as the product. Occasionally, the the original supplier will be willing to sell both the trademark and the production know-how for a minimal price.

When seizing these opportunities, it is important to copy the "old" packaging as closely as possible, so the loyal customers are not confused. Improvements in container and label can be made after the previous distribution pattern has been completely filled. Minor improvements in the food itself can probably be made at the start, provided significant changes in appearance do not result. Avoid any changes that degrade quality or lead to increased prices, at least, until the market is fully adjusted to the changeover.

Legal Pitfalls in Unsolicited Ideas

It is a common belief that vast numbers of ingenious product ideas exist in dormant form among the general consumer population and that this supply could be tapped to the great benefit of the company if only the proper lines of communication could be opened. This belief exists not only among many executives of manufacturing firms, but also among the general consumer population, who ought to know better. As a result, suggestions from outside persons are often encouraged by companies. This is usually a fruitless and sometimes a very expensive policy. Kante (1967) says one company reported that in ten years it had not received a single letter with an idea which could be used successfully.

Most corporate legal officers would agree that, unless unsolicited suggestions are handled with care, the company could be faced with lawsuits from forgotten letter writers every time it puts out a new product. A number of immediate and potential obligations are incurred when a firm receives a new product idea from someone not employed by the firm to create new products. Submittal of an idea by an outsider does not give the firm authority to use the idea in any way other than the way specifically stated by the submitter. The firm does not acquire any property right in the idea by the mere receipt of it, and if the idea is negligently or deliberately disclosed by an agent of the firm to someone who does make use of it, the firm may be liable for damages (Auber 1965). Furthermore, if the firm is in the process of developing or marketing a product similar to that described in the submitted idea, the firm may be put to the expense and risk of proving in court that the concept was actually fully developed by an employee of the firm prior to the firm's receipt of the suggestion.

In a case of this sort was recently litigated (Deogun 1993). A small company developed molding dies consumers could use to make cookie-like confection pieces from puffed cereals and other ingredients. The principals of the firm offered to sell the molds or license the idea to a large cereal manufacturer. Initial management interest was followed by rejection. Then, the cereal manufacturer developed its own line of molds which it offered as a premium to buyers of its products. Their molds had some, perhaps many, similarities in shape, size, etc., to the shaping devices previously offered by the small company. The cereal manufacturer was, of course, sued by the outside inventors. It may very well be that R&D or marketing people in the large company had previously worked on a similar concept which ended up in the inactive files and was resurrected after the outside parties stimulated the manufacturer's interest, but proving this is likely to be difficult. At this writing, the case is still in the courts and may be for some time.

As stated by Crawford (1975), "[If an] . . . idea concerns a new product; it may have value; it is presumably not publicly known; and it is offered in the manner that is not free of conditions even though none may be stated in the letter." Positions taken by the courts over the last 50 years show that it is not legally advisable for a firm to (1) discard unsolicited ideas and take no further action, (2) tell the submitter you already know about his idea, (3) evaluate the idea, or (4) write to get more information about the idea.

If the firm is currently marketing (or has marketed) a product that is exactly and completely like the proposed item, this can be conclusive proof of priority, but the company may still be put to the expense of proving its ownership of the invention in court, a very expensive and time-consuming procedure. Many after-the-fact inventors develop a paranoia about their "rights" that prevents them from accrediting evidence others would consider conclusive, and they will support interminable lawsuits trying to establish their beliefs. In fact, multiple inventors may come out of the woodwork if the invention is a commercial success and gains a great deal of publicity.

According to Crawford (previously cited), most corporate legal departments have adopted procedures to be followed on receipt of a new product idea. These procedures generally include the following steps:

1. If the suggestion is made orally, have the suggester put it in writing.
2. Whoever receives the idea is to send it (including any accompanying documents such as drawings) directly to the legal department. The suggestion is not to be evaluated or even read. If the outside address on the envelope indicates a suggestion (e.g., if the envelope is addressed to the New Product Department), the envelope is not even to be opened until it reaches the legal department. If not forewarned by such clues, the opener of the letter should stop reading it as soon as it becomes clear the letter contains a suggestion.
3. In the legal department, the idea letter is given a number or an identification of some other type and the suggester is sent a form letter that thanks him, tells him the conditions under which the company will consider his suggestion, and (usually) includes a waiver of some type that the suggester is to sign and return. The waiver stipulates, among other things, that the company is obligated only to a minor extent if it uses the suggestion and the company need compensate the submitter only in the amount ultimately agreed to by both parties in a formal contract. Some companies stipulate that the firm may use the idea and pay only what the firm determines to be reasonable compensation, although the enforceability of the latter condition is somewhat doubtful.
4. When and if the waiver is returned by the suggester, the original sug-

gestion is taken from its place of restricted-access storage (e.g., a safe) and sent to the proper party (engineering, R&D, marketing, etc.) for evaluation. Events that follow these reviews are dependent on the perceived value of the suggestion.

5. Along the way: (a) No correspondence acknowledging receipt of the idea is sent to the submitter by any employee or agent outside the authorized department; (b) No one asks for more information from the suggester until a waiver is received; and (c) Even if the employee receiving the suggestion thinks it is obviously not new or is completely lacking in merit, it is handled in the prescribed manner.

MARKETING DEPARTMENT'S ROLE IN IDEA GENERATION

In addition to participating in the overall operation of systems set up for the generation of ideas, marketing personnel can play a useful role in the company's new product development program by identifying specific unmet needs of consumers. Marketing's contribution should take the form of a well-organized, carefully planned, and expertly performed program of consumer research, including focus group discussions and in-depth interviewing (Moldawer 1976). The goal should be to ferret out those consumer desires that are poorly visualized and imperfectly formulated by most or all consumers but which can be satisfied by a product that is technologically feasible. Since this approach to focusing idea generation requires imagination, careful planning, and a great deal of work, as well as considerable self discipline, it is rarely, if ever, taken by small companies. It could be one of the most rewarding methods of guiding product development, however.

Consumer research of the type described in the preceding paragraph should be designed to target a specific consumer population. A decision can be made to appeal either to the mass market (the majority of the consuming public) or to some minor segment of the population that has special, possibly bizarre, requirements and tastes. If groups of the latter type are targeted, distribution and marketing will need programs that are much more creative and insightful than usual, but profit margins and customer loyalty are likely to be greater, and advertising and promotional expenses may be less because of limited competition.

Technical product development for special groups need not be more difficult than for mass market items and may actually become relatively simple once the peculiar requirements of the targeted population are completely understood, because the limited choice available to such customers may cause them to be less critical of minor product deficiencies. The health food market is an obvious example; various ethnic groups also constitute possible specialized markets.

In commercial foods, there are many situations where contrasting approaches, mass market vs. specialized market products, have been successful. Table 2.1 gives a few examples.

Table 2.1

COMMERCIALLY SUCCESSFUL FOODS HAVING CONTRASTING APPEAL

Mass	Specialized
Processed American cheese	Limburger cheese
Catsup	Tabasco
Wonder bread	Pepperidge Farms bread
Milk	Buttermilk
Budweiser beer	Champale
Potatoes	Broccoli

It appears that previously unknown market segments (or potential segments) can sometimes be recognized by studying the distribution of hedonic scores obtained from taste tests of food products that have average scores in or near the dislike range but also have an appreciable number of individual ratings at the high end of the scale. If the responses when tabulated or plotted clearly show a dichotomous distribution, with a majority of the ratings in the dislike range, but a substantial number of ratings at the "Like very much" or "Like extremely" levels, an opportunity may exist for appealing to a specialized group of potential customers.

After establishing presumably rather diffuse limits of the desired product characteristics by a study of consumer research results, it would be appropriate to request R&D personnel to describe ways in which some or all of these characteristics can be achieved. Of course, some of the desired qualities may be inconsistent with others, or physically unachievable. Still others may not be permitted by federal regulations or company policies. The general process for analyzing the practical feasibility of a concept can follow a path such as that described by Schrello (1975). When a concept can be drawn up incorporating all of the desired characteristics which can be achieved technically, it should be submitted to consumer panels to determine whether it adequately meets the unfilled need or has been modified so much as to be unacceptable. Further development can be guided by one of the systems to be described in a subsequent section.

RECORDING IDEAS

It is essential that a register be established for all ideas that enter into the collection and evaluation procedure of a company. This register, which may take the form of a collection of filled-in forms, must be under the supervision of one (only) authorized person, who need not be a technically trained person, since the duties are largely administrative in nature. Each idea, as represented by a single form, must be given a number that will identify it through all subsequent operation. The administrator should keep the numbers, with a brief description or expanded title, in a bound book dedicated to this specific purpose.

Figure 2.1 gives an example of a form that might be used. It should be adapted to conform to the needs and customs of your company.

IS A MORE SELECTIVE APPROACH JUSTIFIED?

Many of the techniques that have been suggested as methods for generating or acquiring ideas seem to be primarily concerned with merely increasing the number of ideas, i.e., the more new concepts that can be recorded, the better. It could be argued that this is the wrong approach, that the time and effort required for skilled personnel to apply the necessary screening and evaluation procedures could be greatly reduced if idea generation was focused on certain predetermined goals at the outset.

A director of the commercial development division of a large corporation said, "We accept few ideas for evaluation . . . because our experience has indicated the cost of evaluating available ideas far outweighs potential benefits in new product development . . . Outsiders will, no doubt, come up with many good inventions which we miss. Even so, we have found that individual inventors and the public at large vastly overrate the value of an invention. There is a great deal of cost, risk, and effort required to make a device or technology into a product which will be marketable . . . Only rarely and with great effort do good ideas become successful products" (Anon. 1978). Most of us could agree with the general tenor of these observations while at the same time suggesting that an efficient method of screening would eliminate some of the problems anticipated by the commentator.

Focused idea generation, which would eliminate the entry of many unsuitable ideas into the system, could be based on the prior establishment of parameters describing the kind of ideas that are needed by the company. This is the type of idea generation used in the most efficient scientific experimentation, for example.

However, since disciplined idea generation does require more mental effort, more self-discipline, more familiarity with the subject, and (usually) greater intelligence, it will always be less popular than more relaxed, "fun" type techniques such as brainstorming. As stated by Moore (1971), "From the human viewpoint, the fun of creating is replaced by the work of difficult decision-making. For the group, there is a clear change of mood from the certainty of joint creativity to the uncertainty of mutual agreement."

In any case, at some time during a product development program a weeding-out process must be applied to the stock of ideas, no matter what technique is used to generate those ideas. The following chapter is devoted to a discussion of methods that have been recommended for culling out the less promising ideas.

Idea number:______________

1. SOURCE OF IDEA:
   Name of submitter:____________Date received____________
   Is submitter an employee?____ Title: ____________________
   Home address of submitter ______________________________
   ________________________________Phone ____________

   Form in which idea was submitted (written, oral, model):
   ____
   Does submitter claim the idea originated with him/her?:
   ____
   If not, where was idea obtained? ____________________
   ____

   Has submitter ever tried to put the idea into practice?:
   ____
   Has submitter offered idea to another firm or person?:
   ____

2. DESCRIPTION OF THE IDEA:
   General class: Food, beverage, container, process, or?:
   ____
   Narrative description of the idea (as nearly as possible, in the submitter's own words):________________________
   ____
   ____
   ____
   Name an existing object or process the idea resembles:
   ____

3. PRELIMINARY EVALUATION OF THE IDEA:
   How does the idea improve on existing art?:______________
   ____
   Does our company make or use an existing counterpart?:
   ____
   Has anyone in our company previously expressed a need for such an invention?:______________________________

Recorder:
Name:__________________
Title:__________________
Date:__________________

Witness:
Name:__________________
Title:__________________
Date:__________________

4. DISPOSITION OF THIS IDEA (initial and date each action):
   a. Put on inactive status? Why? ____________________
   b. Evaluation form forwarded ______________________
   c. Submitter notified of action ______________________

Figure 2.1. Register of ideas for a small company.

BIBLIOGRAPHY

ANON. 1967. How to generate ideas for new products. Business Management *33*, No. 1, 83-91.

ANON. 1978. New ideas needn't be a nuisance. Industry Week *197*, No. 4, 106-189.

ANON. 1989. Unlocking creativity in engineering. Designfax *11*, No. 4, 30, 32, 37-40, 42

AUBER, R. P. 1965. Outside ideas – dynamite! Res. Management *8*, 183-190.

BARTOW, P. E. 1975. Brainstorming. Management Rev. *64*, No. 11, 54-58.

COBB, E. 1986. Learning the lessons of R&D. American Heritage of Invention and Technology *1*, No. 3, 8-9.

CRAWFORD, C. M. 1975. Unsolicited product ideas – handle with care. Management Rev. *64*, No 8, 54-57.

DETWILER, R. M. 1976. Madison Ave.'s one-time mad fad. Advertising Age *47*, No. 38, 59-60.

KANTE, W. 1967. Unsolicited suggestions: Handle with care. Management Rev. *56*, No. 10, 36-38.

LAW-YONE, W. 1980. Company Information: A Model Investigation. Washington Researchers, Washington, DC

LEVY, P. 1993. The granddaddy of food gatherings. Wall Stree J. *92*, No. 14, A-12.

MALOVANY, D. 1990. The search for new ideas. Bakery *25*, No. 6, 43-44,46.

MARLOW, P. 1987. Qualitative research as a tool for product development. Food Technol. *41*, No. 11, 74, 76, 78.

MOSKAL, B. S. 1977. Sleuthing the opposition. Industry Week *195*, No. 4, 52-57.

TAUBER, E. M. 1971. Systematic generation of ideas for new foods. Food Prod. Dev. *5*, No. 2, 58-59, 62.

TAYLOR, D. W., PERRY, P. C., and BLOCK, G. H. 1958. Does group participation when using brainstorming facilitate or inhibit creative thinking. Admin. Sci. Q. *3*, 23-47.

chapter three

SCREENING AND EVALUATING IDEAS

INTRODUCTION

Any company making a concerted effort to generate and collect new product ideas will soon accumulate a stock of concepts exceeding the number that can be worked on by their product development department. Experience has shown that most of the ideas obtained by any method will be of little or no use to the company. For these reasons, and others, a system must be established to screen each idea that is submitted, not necessarily to evaluate all of them in depth, but at least to scrutinize each one with sufficient intensity to insure that important opportunities will not be missed and that additional time will not be wasted on clearly unsuitable concepts. Candidates having the least promise must be eliminated from additional consideration, while those seeming to be both feasible and profitable will be entered into the project flow and subjected to further analysis and development.

It is important that each idea be graded by the same standards during the screening process. This implies that an idea evaluation policy must be in place, and that its procedures must be followed consistently. Responsibility for doing the screening must be clearly assigned to a named person or to a small group having the necessary qualifications, and an evaluation procedure must be set down in writing and made known to the assigned authorities. It is often helpful to design a form or set of forms upon which can be recorded the idea and the results of its evaluation, together with reasons for the decisions that were reached.

Even though every idea should be judged by the same standards, the screening policy should allow prompt rejection of certain ideas. A suggestion that a food company manufacture a perpetual motion machine does not require a great deal of deliberation before it is rejected. To spend as much time on such a concept as on a fully described proposal for, e.g., a new type of dietetic chocolate bar would be a waste of valuable technical skills.

Prioritizing is not normally part of the screening process; that comes later. In one sense, though, it might be said that the object of screening is to give each idea a priority of 0 or 1, fail or pass.

The screening process does not result in the immediate establishment of a new product development project. The critical steps of evaluating an idea's potential worth and determining how the product it describes can be integrated into the company's long-term plans must be performed before the idea can be made into a project. An idea approved in the initial evaluation process must be further refined to yield tentative product specifications that are incorporated into a project plan containing deadlines and other scheduling features as well as funding details. A project proposal will also have a priority assigned to it at some time, whether at the time it is established or later by a separate priority committee. The procedures for drawing up project proposals and setting priorities will be described in subsequent chapters.

Acceptance or rejection of an idea should take into consideration the overall mission of the company, with careful attention given to the resources available for performing the work needed at each step of the process. To determine whether or not a product idea conforms to the needs of the company, it is important to have in place a company Mission Statement.

MISSION STATEMENTS

In order to establish any kind of realistic plan for progress, a company must know what it is doing at present, what resources it has available, and where it wants to go in the future. A clear understanding of these points, as they are embodied in a Mission Statement, will provide a framework for the system by which new product ideas are screened and evaluated. Unfortunately, the top managements of many companies do not have a good grasp of these basics, being more concerned with everyday problems of finance, marketing, personnel, public relations, and government regulations. Where their company will be in five years is likely to be regarded as being less important than the price their stock will be selling for next week.

In addition to its other justifications, preparing a Company Mission Statement is an initial step in clarifying the firm's basis for judging the suitability of new product ideas that have entered the system. It sets the basic outline for the criteria the new product development manager will establish for deciding the economic desirability of an idea; there will be other criteria (such as technical feasibility) that have little or nothing to do with the Mission Statement.

Drawing up a Mission Statement is neither very difficult nor time-consuming if its proper role and function in company management strategy are understood at the outset. A "mission" is a general, not a detailed, plan for the future; in the best cases, it is relatively short on itemization. The intention is not to place unnecessary restrictions on either present operations or future expansions of the company, but to map out the general direction the company expects to take in its progress toward becoming a more efficient and profitable organization.

Many companies — including some very large ones — issue "Missions" that are largely public relations ploys, and very little else. Sometimes, they are called "Visions." General Motors issued the following "Saturn Mission:"

"Market vehicles developed in the United States that are world leaders in quality, cost and customer satisfaction through the integration of people, technology and business systems and to transfer knowledge, technology and experience throughout General Motors."

Similarly, Levi Strauss & Company issued a high-flown "Mission Statement" seeming to recognize the effect of divine intervention on the peddling of jeans:

"The mission of Levis Strauss is to maintain responsible commercial success as a global marketing company of branded casual apparel. We must balance goals of superior profitability and return on investment, leadership market positions, and superior products and service. We will conduct our business ethically and demonstrate leadership in satisfying our responsibilities to our communities and to society. Our work environment will be safe and productive and characterized by fair treatment, teamwork, open communications, personal accountability and opportunities for growth and development."

Statements of the foregoing type offer no help at all to the struggling new product group.

It is not necessary for the authors to detail the reasons for each point they set down in the Mission Statement, since it is assumed that a CEO and his advisers have adequate reasons for making the decisions and the authority for making them. A Mission Statement is not a document the justification for which is intended to be debated at each management meeting.

One of the basic purposes of a company's Mission Statement is to publicly inform its stockholders and creditors as well as its employees of the limits within which the company will perform its activities and the goals toward which the company will direct its energies. It will take into account the resources available, or to become available, to the company and all the forseeable restrictions on its actions.

Procedures for Preparing Mission Statements

Although the CEO is the nominal author of the Mission Statement, it is quite likely he will assign its responsibility for its composition to other persons. It is highly desirable that the authors of the Mission Statement have a clear view of the company's long-range goals as well as a deep familiarity with the present nature of the company and the way in which it conducts its business. Consultants charging up to $3,000. per day (Fuchsberg 1994) are available to help if the CEO develops writer's block.

The statement should be broad enough to encompass all activities into which the company plans to enter. It should avoid excessive optimism and banal generalities, and it should not include vague or overly broad guidelines. The best Mission Statements give down to earth guidance to people who work for the company, allowing them to control their activities in a way most useful to their employer, and will, as well, allow them to know how they will be treated by the company.

It might be worthwhile to include in the Statement a few words defining how the company regards its relationship and obligations to society as a whole, as long as the declaration is believable and realistic. On the other hand, preachments, philosophical abstractions, and grandiose posturings do not belong in a Mission Statement — it should be neither a press release nor a wish list. Don't expect to gain any points with your employees, the public, the press, or activist groups by including idealistic pronouncements that can't be translated into actions — you lay yourself open to disparagement or even ridicule when the promises turn out to be impossible of attainment.

Various members of management should be asked to review the Mission Statement and make suggestions for revisions, but the final judgement as to the form and content of the Mission must remain the privilege of the CEO and the Board of Directors since they have the ultimate responsibility for directing the company.

Usually, the total company will have only one Mission; if necessary, the Mission may have several parts. Divisions and Subdivisions of the company may be required to prepare their own Mission statements, but it is probably more appropriate to allow the separate departments to draw up sets of Objectives, which serve somewhat the same purpose as Mission statements but are more detailed and are, as a rule, subject to more frequent changes.

Although a Mission is not to be regarded as a ephemeral document, subject to change day by day, it is very obvious that alterations in its content may be made necessary by the march of events and by occurrences that could not have been foreseen at the time of its inception. Changes should be regarded as serious, not trivial, occurrences, and should be decided upon only after consultations with all executives who might be affected by the modifications. When changes are put into effect, they should be made known to all persons familiar with the preceding version.

Guidelines for Composing Mission Statments

Several consultants and administrative specialists have published articles describing the procedure of writing a Mission Statement and listing the items it should contain. According to Rives (1964), somewhat modified here, top management must take the following steps:

(1) Define the nature and acceptable boundaries of the business.

(2) Establish long-range business goals that present attainable challenges.

(3) Provide resources adequate for achieving to the assigned objectives.

(4) Insist on goal-oriented planning by functional operating groups to ensure valid management decisions.

(5) Establish channels of communication.

(6) Organize so the contributing functional groups must communicate.

(7) Make sequential decisions consistent with the long-term objectives, but subject to correction according to current input of changing data.

(8) Monitor progress.

MISSION STATEMENT
of the
SUPER SCOTCH BAGEL COMPANY

1. Our principal mission is to bring to consumers a wide choice of healthful foods that look, taste, and feel good, and that can be purchased at or below competitive prices. In accomplishing this goal, we will adhere to the following guidelines.

2. Our loyal employees will be provided with a safe and congenial workplace and with fair and equitable rules allowing them to function without harassment or undue restrictions on their activities. Their compensation will be equal to, or better than, salaries paid for equivalent jobs by other employers in the same business.

3. We recognize our obligations to the company's owners and will conduct our business in an efficient and cost-conscious manner so that shareholders can realize a fair return on their investment while at the same time having a reasonable expectation of increases in the share price.

4. Our financial reports will be timely, accurate, and complete; they will be compiled by accepted accounting methods and will be made available to the public.

5. While at all times obeying the laws and regulations of each jurisdiction within which we operate, we will oppose by all legal and ethical methods those restrictions and exactions that we consider unfair and impractical and which place undue burdens on our industry.

6. We will adapt quickly to changing consumer demands, and will attempt to anticipate these trends. In pursuance of this policy, we will maintain an active and capable product development group.

7. We will support the independence of our quality assurance function, to more completely ensure the safety, uniformity, and regulatory conformity of all our products

Figure 3.1. Mission Statement of a fictitious company.

Another view of the composing of company mission statements is attributed to Dr. R. E. Linneman, who listed ten steps to be followed:

(1) Identify and make explicit your company's basic objectives and policies.

(2) Determine how far into the future you wish to plan.

(3) Develop a good understanding of your company's points of leverage and vulnerability.

(4) Determine factors you think will definitely occur within your planning time frame.

(5) Make a list of key variables that will have make-or-break consequences for your company.

(6) Assign reasonable values to each key variable.

(7) Build scenarios in which your company may operate.

(8) Develop a strategy for each scenario that will most likely result in achieving your company's objectives.

(9) Check the flexibility of each strategy in each scenario by testing its effectiveness in the other scenarios.

(10) Select or develop an optimum response strategy.

An example of a Mission Statement that might be adapted for the use of a company engaged in the food manufacturing business can be found in Figure 3.1.

OBJECTIVES OF TECHNICAL GROUPS

Every company embarking on a formal product development program or formalizing its current program should supply a copy of the Mission Statement to the person who has charge of the technical operations of the new product function, such as the head of the Research and Development Department or the manager of the Product Development group. The laboratory director can then draw up a set of Objectives applicable to the personnel under his control and conforming to the company's Mission. Of course, these sets of Objectives must be approved by, or drawn up in conjunction with, top management.

It may then be considered desirable to compose a set of Objectives for the product development group, possibly as a sub-set of the R&D Department's Objectives. The product development statement will consist for the most part of detailed and specific directives deriving from the Mission Statement, and they should be drawn up in conference with the laboratory director and those technical people who can provide valuable input into the process.

Fundamental considerations, When corporate management draws up statements of mission or objectives applicable to product development operations, they must understand what they want to achieve from new products. The corporation's strengths should be evaluated in order to determine the general type of products on which to concentrate and the level of support that can be given to research and development. Objectives should take into account factors such as corporate image, existing technology, ability to procure raw materials at reasonable prices and in required quantities, manufacturing capabilities, type and structure of distribution facilities, and focus of marketing expertise (the preceding discussion was based partly on Sullivan 1976).

Desrosier and Desrosier (1971) quoted a typical research laboratory mission statement, as follows:

1. Develop and keep abreast of scientific and technical knowledge in areas of vested corporate interests.
2. Apply this technical knowledge by developing new and improved products and processes for the corporation.
3. Establish, issue, and maintain standards and specifications for raw materials and finished packaged products bought and/or sold by the corporation.
4. Audit adherence to the approved quality standards and recommend corrective action as required.
5. Represent the corporation in matters involving laboratory sciences and technologies.

Some managers will regard this list of duties as being too specific for a statement of mission, which should be brief, general, broad, and relatively timeless. A mission statement that would be appropriate for most companies would be: "The Research and Development Department shall utilize the principles of science, technology, and engineering to expeditiously and economically develop new and improved processes and products that are compatible with the resources and long-term goals of the company."

Examples of objectives. According to the views of Heid (1963), the objectives of applied research and development activities in the food industry are:

1. To produce, maintain, and increase sales by:
   a. Improving processes, products, and packages.
   b. Developing new uses for products.
   c. Developing new products, processes, and packages, plus testing acceptability of these products.
2. To eliminate losses and hazards of losses by troubleshooting and by developing profitable utilization of surpluses, residues, and wastes, and by avoiding nuisances, damage suits, etc.
3. To minimize dependence on raw materials which are in inadequate or undependable supply.
4. To act as technical consultants to top management in evaluating proposed new equipment and processes, keeping management informed of developments of potential interest and importance.
5. To increase the return on invested capital and profit on unit sales volume by insuring better, more uniform products at the lowest possible

cost.

6. To deal with government agencies in relation to technical matters including labeling, standards of identity and quality, waste disposal, etc.

Other authors would undoubtedly regard some of these functions as being more appropriate for a QA/QC or technical service department.

CHOOSING IDEAS THAT MEET COMPANY GOALS

A product development program should have as its primary goal the improvement of the company's financial results by introducing into the marketplace new products that attain good consumer acceptance and return a satisfactory profit. It is generally expected that total sales volume will also increase, but this need not always be the case (e.g., when an unprofitable item is replaced by a profitable one). A good product development procedure will utilize in the most efficient way all of the technical, marketing, manufacturing, sales, and financial resources of a company. It must prevent the initiation of projects that will ultimately call for resources beyond the company's capabilities. It should take into consideration the company's long term policies as well as the image it wishes to present to the consumer.

The procedure should provide for constant review of the desired product characteristics in the light of changing federal and local laws and regulations. Although it should not allow risks to be ignored, it should include provisions for accepting certain risks if potential rewards are sufficiently great. Responsibilities should be unambiguously assigned by the procedure, and authority should be clearly delegated. It should make it clear that stalemates and other avoidable delays are unacceptable.

The plan should allow participation in the development process by all departments of the company that can be expected to make worthwhile contributions, but it should not involve extraneous persons who would only complicate the efforts of productive workers and thereby delay completion of the project. Some streamlining of the staffing and information flow may be necessary to achieve expeditious accomplishment of the mission, even though it may add to the risk of some errors being made. The first step in the plan is accumulating or originating ideas, the second step is matching the new products that would result from these ideas with the company's mission, resources, and objectives.

Defining the Fields of Interest

After agreement has been reached on a Mission and Objectives, management has a basis for drawing up a plan describing in considerable detail its new product interests for the future. These areas of interest should be made as narrow as possible to avoid wasted effort. A statement such as, "This company is interested in any product that will make a profit," is evidence of an abdication of top management's responsibility to provide necessary guidance. The chief executive, or his delegate, should be aware of the company's limitations in marketing capabilities, capital availability, legal restrictions, plant capacity, and other factors affecting the type of product the company can make successfully and distribute profitably, and these limitations should be communicated to the persons responsible for new product planning. The more clearly these limitations are set forth, the less is the chance that time and resources will be spent on unsuitable projects. The first and possibly the most crucial evaluation step is the screening of ideas.

Within the fields of interest which have been established, marketing research should be asked to identify rather broad areas of marketing opportunities. Generally speaking, it is of doubtful value to ask market research to specify a definite product. Desires of the consumer are likely to be rather amorphous until crystallized by the offer of a product they can see, hold, feel, or taste.

Furthermore, specifications arrived at by consumer research may be impossible of achievement in practice. A better approach is to design a product on the basis of broad guidelines developed from market research studies.

If a superior or novel product becomes possible of achievement because of advances in technology, it is perfectly in order for R&D to develop a prototype and then request marketing to determine consumer acceptability of that specific item.

Examples of Systems for Screening Ideas

Ellington (reported by Marting 1964) also gave suggestions for the design of a form on which new product ideas can be submitted. The most prominent features of his system are:

1. Ideas should flow to one designated point of contact.
2. All submittals should be acknowledged within a reasonable time span, and a second communication should follow after action has been taken. If the product idea does not appear acceptable to those responsible for development, logical reasons should be given for rejection.
3. An award plan, if any, should be clearly described. Those persons ineligible to benefit from the plan, such as top management and those persons specifically employed to develop new products, should be clearly defined.
4. The decision as to worth of ideas submitted should not be left to the judgment of one individual. At least once a month all ideas submitted since the last review should be considered by a New Product Development Committee. Ideally, this committee should be made up of representatives from sales, manufacturing, and engineering, as well as from market research, the comptroller, and new products, if the company has these functionaries. Specific reasons for rejection of ideas by the committee should be recorded. If the committee feels the idea has merit, it should undergo a period of

evaluation, which might include a market study, engineering design work, cost estimation, or other activities, before a decision can be reached that the idea warrants a full-scale development program.

A list of questions to be used for evaluating the suitability of new product ideas was suggested by Glasser (1982). It is reproduced below in shortened form.

1. What problem are we trying to solve?
2. What is it we must bring to the marketplace?
3. Who can buy?
4. What is the market timing?
5. Who is our competition, now and in the future?
6. What constitutes the incentive to buy?
7. What are the key milestones to acceptance? [advantages over competition].
8. Is our solution compatible with current technology?
9. What type of selling is needed?
10. What are the legal constraints [safety, pollution, etc.].
11. What is forcing the development of it?
12. Is there a requirement or do we create one?

Glasser points out that "[other] specific questions should be developed for . . . specific programs."

During the 1970s, Dr. Gerald Udell (then associate professor of marketing and director of the Experimental Center for Advancement of Invention and Innovation at the University of Oregon) developed a 33-point model for examining potential new products in terms of costs, payoffs, and the innovation strategies required, rating them in these factors from very low to very high. Five overall categories in his system were:

(1) Societal factors. This seems to relate to an evaluation of the idea from an ethical and legal standpoint, with safety, environmental effects, and the like entering into the equation.

(2) Business risk factors, which force the innovator to confront many risks by comparing his product with similar competing items already being marketed. Even though the product has a high degree of functional feasibility, production feasibility, and investment cost, there may be insufficient profitability, a payback period that is too long, and market research that is too costly and unreliable.

(3) Demand analysis factors. How many units can be sold in the local or national market that is targeted? How will the market hold up; repeat sales, etc.

(4) Market acceptance factors try to define how the existing market will accommodate the new item. The proposed product may be undesirable if consumer attitudes are indifferent or hostile, if the time and money required for promotion, consumer education, and the creation of new distribution channels are too expensive and too time consuming.

(5) Will the product's appearance, function, durability, and price be

attractive enough to lure consumers away from already existing products?

The current status of Dr. Udell's system is not known, but the program does appear to be proprietary, in its details and application, so he would be the source of further information.

Lauro (1977) recommended a two-tier method of screening ideas that takes into account numerous regulatory and nutritional considerations. Perhaps most managers would consider the list too extensive for the level of idea evaluation being discussed in the present chapter. Lauro stated the older method of screening might involve:

(1) Does the product idea fit the corporate line of products?
(2) Does the product idea satisfy a consumer need?
(3) Is the market segment large enough to build a business on?
(4) Do we have the resources to counter a competitive response?
(5) How much will it take to develop and introduce?

Newer conditions affecting the decision to commence development work would include (abbreviated here): (1) Will the product contribute substantially to the diet? (2) Are the ingredients most likely to be safe? (3) Do we have suitable GMP facilities for manufacture? (4) Will waste and pollutants result? (5) What kind of claims will Marketing want? (6) Is there a suitable package available? (7) What will be the associated capital cost needed to prove adequacy of the process? (8) Will the product cause a consumer hazard in distribution? (9) What is the cost of QC? and (10) Will the consumer encounter a hazard in using the product?

Practical Considerations in Screening

One of the first goals of the screening process must be to decide how an idea fits company objectives. An idea can be superb for one company but totally unsuited for another. There should be very few, if any, situations requiring the modification of company objectives to fit the needs of an idea; the intent is to determine whether or not the proposed new product can be developed and marketed by the company within the constraints imposed by its current image, goals, and resources.

The practicality of an idea depends to a large extent on its fit with the company's existing strengths and weaknesses; it must describe a product that can benefit from the strengths and that will not be seriously affected by the weaknesses. It follows that these factors must be known in order to have a sound basis for judging the value of an idea. An initial step in determining the strengths and weaknesses will be the making of an inventory of the company's capabilities and limitations. This list will ordinarily be much more detailed and quantitative than a Mission Statement. Some of the items that should be included in the list are:

1. Would the product be compatible with the image the company wishes to project?
2. Would the product be in accord with company policies?

3. Is the present marketing staff capable of handling a product of this type?

4. Are similar market niches occupied by present product lines?

5. Are existing methods of distribution suitable for the proposed product?

6. Are the company's technical strengths sufficient for developing such an item?

7. Can the product be made wholly or partially on existing production lines?

8. Can financial resources be made available for all phases of the development process?

9. Are there legal limitations preventing the merchandising of the product (such as restraint of trade considerations, etc.).

10. Geographical considerations: supply availability, location of most probable markets, etc.

The smaller the organization, the more stringent must be the selection process in order to match the limited time available for prototype development with the requirements of the company for new products. As a practical matter, there should be a tendency to err on the side of rejection rather than on acceptance, because of the unexpectedly large demands on company technological resources that always occur when a new project is started.

At times, the rejection of a non-conforming idea requires considerable restraint and self-discipline on the part of executives, since some of the unsuitable ideas may seem to be so outstanding in their degree of novelty and consumer appeal that there is a strong temptation to force them into the new product development program in spite of a clear recognition of their incongruities relative to company policy. In such cases, it may be possible either to set up a new venture group having the sole function of bringing the new product to market outside of regular company channels (a very expensive process) or to market the idea to another manufacturer.

How Many Categories?

The result of the work of the screening person or committee will be the the placing of an idea into one of two or more categories. The simplest (and in this author's opinion, the best) scheme is to classify the ideas as "Suitable for further work" or "Not suitable." In this dichotomous arrangement, the rejected items may be temporarily or permanently discarded, depending on the method selected – the treatment should be consistent for all ideas, however. In general, it would appear best to completely remove an idea from the stock once it has been rejected. The recycling or reviewing of cast-off concepts takes time that could be better employed in reviewing new submittals. Accepted ideas progress into the project establishing stage where still further evaluation and refinement of the concept takes place, with the ultimate goal of developing a marketable product.

Some authorities do not agree that a simple "yes or no" scheme is best.

For example, Souder (1984) stated, "In screening, the idea is analyzed in a preliminary fashion, on the basis of its most prominent criteria or characteristics. The screening process is a kind of quick and inexpensive first level analysis of the idea. The results from the screening process may suggest that the idea should be rejected because it is not meritorious . . . (Abandon idea). Or the idea may be temporarily backlogged (Idea backlog) . . . in deference to a higher priority idea. Or the idea may be deemed so urgent (Urgent item) . . . or so obviously beneficial to the organization that it is accorded priority over all other efforts. Alternatively, the results of the screening analysis may indicate that the idea warrants further investigation and a more in-depth proposal." I see nothing particularly wrong with this scheme, except that it complicates the product development process at a point where, in food product development at least, rapid progress and simplification are most needed.

If management feels uncomfortable with the idea of casting new product ideas into the wastebasket, a less painful alternative for the idea evaluation group may be to use three or four categories, such as:

(1) Accepted ideas, to go immediately into the product development flow with the establishment of a project as the next step. The active file.

(2) Extremely urgent projects, considered to be of utmost importance to the company's survival; these would be transferred to the project establishment file with an automatic priority of "1," or another priority as decided by the CEO. It is to be hoped that such projects would be rarely encountered, otherwise they will, in short order, completely overwhelm the normal course of new product development.

(3) A "waiting at the end of the line" category for those ideas deemed to be worth further consideration but which are, for some reason, not suitable for a complete evaluation at the time. The inactive file.

(4) A category for those ideas considered to be of no value at the time of review or at any conceivable point in the future. They will be withdrawn from the system. Rejects.

RECORDING THE EVALUATIONS

It is very helpful to have a standard form upon which the various interested parties can enter their comments as to the suitability of an idea for their particular purposes. The completed evaluation forms should be kept under the control of a single person, not necessarily an executive. A competent and reliable secretary is often a good choice for maintaining this file. An example of an idea evaluation form suitable for small companies is shown in Figure 3.2. The details will undoubtedly have to be modified to meet the needs and organization of your firm.

Idea number:______________

1. FROM: __________________ Date forwarded__________________

   TO: Director of Marketing
   Production Manager
   Legal Department

   We submit the following idea to you for an evaluation based on the requirements of your department. R&D briefly reviewed the concept and reported there are no serious technical problems involved in its implementation. Consider all factors that might affect the success of this [product/package/process] if it could be developed.

2. DESCRIPTION OF THE IDEA:
   General class: Food, beverage, container, process, or?:
   ____

   Narrative description of the idea as interpreted by the R&D department: ____________________________
   ____
   ____

   R&D comments on possible positive and negative features:
   Positive:_____________________________________
   ____
   Negative:____________________________________
   ____

3. EVALUATION OF YOUR DEPARTMENT (Attach sheets if needed):
   Marketing evaluation (does a need exist?):_______________
   ____
   ____

   Production evaluation (can this be made with existing equipment?):_________________________________
   ____
   ____

   Legal evaluation (can any regulatory problems be foreseen?)
   ____
   ____

   Should other departments be asked to review this idea? If so, state who and why:______________________________
   ____

   Please complete this preliminary review within 5 working days of receipt, then immediately forward the form to the next department.
   RETURN TO:

4. COORDINATOR'S ACTION: (Initial and date when action taken)
   (1) Placed on inactive list ______________________
   (2) Forwarded to CEO for approval of Project Proposal __________
   (3) Other action (Describe) ______________________

Figure 3.2. Preliminary idea evaluation form for a small company.

BIBLIOGRAPHY

ANON. 1955. Developing and Selling New Products, 2nd Edition. U. S. Chamber of Commerce, Washington, DC

ANON. 1976. From concept to product – in 21 steps. Plastics Eng. *32*, No. 7, 1-4.

BEST, D. 1986. Assessing the risks of research and development. Prepared Foods *155*, No. 10, 150-152, 156.

BONE, B. 1987. The importance of consumer language in developing product concepts. Food Technol. *41*, No.11, 58, 60, 86.

CROCKETT, J. G. 1969. The new product manager in product development. Food Technol. *23*, 26-30.

DESROSIER, N. W. and DESROSIER, J. N. 1971. Economics of New Food Product Development. AVI Publishing Co., Westport, CT

FUCHSBERG, G. 1994. "Visioning" missions becomes its own mission. *93*, No. 5, B-1, B-4.

GLASSER, A. 1982. Research and Development Management. Prentice-Hall, NYC

HEID, J. L. 1963. Research and development. *In* Food Processing Operations, Vol. 1. M. A. Joslyn and J. L. Heid, Editors. AVI Publishing Co., Westport, CT

HILTON, P. 1961. Handbook of New Product Development. Prentice-Hall, Englewood Cliffs, NJ

KOTLER, P. 1967. Marketing Management. Prentice Hall, Englewood Cliffs, NJ

LAURO, G. 1977. Product development today. Food Technology *31*, No. 7, 73-74, 82.

MARTING, E. 1964. New Products – New Profits. American Management Assoc., NYC

MARVIN, P. 1972. Product Planning Simplified. American Management Assoc., NYC

MATTSON, P., and LOTT, D. 1971. What do users and prospective users think of external product development services. Food Prod. Dev. *5*, No. 5, 50, 52.

MOLDAWER, R. A. 1976. The marketing research steps that aid sound product development. Package Dev. Syst. *6*, No. 5, 40-42.

MOORE, L. B. 1971. The problem-solving process. Evaluation. Dream a Little Library, Installment *10*. Atlas Chemical Industries, Wilmington, DE

RIVES, R. C. 1964. What is your new product policy? *In* New Products – New Profits, E. Marting (Editor). American Management Association, NYC

SCHRELLO, D. M. 1975. Twenty questions for product decisions. Industrial Res. *17*, No. 9, 83-84.

SINKI, G. S. 1988. Finding the universally acceptable taste. Food Technol. *42*, No. 7, 90-93.

SOUDER, W. E. 1984. Project Selection and Economic Appraisal. Van Nostrand Reinhold, NYC

STANLEY, A. O. and WHITE, K. K. 1965. Organizing the R&D Function. American Management Association, NYC

SULLIVAN, D. 1976. Good product concepts embody business goals, market needs. Food Prod. Dev. *10*, No. 4, 28, 30.

WALTERS, J. E. 1965. Research Management: Principles and Practice. Spartan Books, NYC

WHITE, V. P. 1988. Handbook of Research Laboratory Management. ISI Press, Philadelphia, PA

chapter four

ESTABLISHING PROJECTS

INTRODUCTION

After it has been decided that an idea has sufficient merit to justify its further consideration as a possible element in the product development program of your company, the next step is to set up a project plan that will include product specifications, a time table, budget details, and (probably) designation of a team to conduct the work. Many more details may be included in a project description, but these are considered by most students in this field to be essential.

Not all approved ideas lead to finished projects or even to completed project proposals. It may very well happen that, in the course of assembling the information necessary to compose the project proposal, serious problems will be uncovered. If this occurs, and the problems appear to be intractable (at least to efforts commensurate with the expected value of the product), there may be little or no technical work done, no project implemented, and the idea abandoned.

According to Bass (1967), the natural sequence of steps in a technical undertaking is:

(1) Searching for a new idea within the field of interest.

(2) Examining the merits of the idea by some kind of evaluation – initially by mental processes and then, often, by preliminary experimentation.

(3) Setting a tentative goal if the conclusions are favorable.

(4) Making a plan that can be used to achieve the objective.

(5) Adopting criteria that can be used to evaluate success of the project.

(6) Undertaking work and accumulating results.

(7) Concluding the project when analysis reveals that the goal has been achieved. The goal may have to be redefined many times before success is finally attained.

The first two steps have been discussed in preceding chapters. The last five, as well as other points not stated by Bass, will be covered in this and subsequent chapters.

Later publications with lists of steps, stages, phases, or points that should be considered during a new product project have become available, but the Bass compilation is still worthy of consideration.

Media attention has recently been concentrated on the use of product development "teams" working on parallel projects, with allegedly great benefits resulting therefrom. The newness and prospective benefits of the parallel team approach may be debatable, but the general idea evidently has some advantages for specific situations. Himmelfarb (1993) listed ten basic steps that should be taken before teams begin work on major parallel development projects. These steps are (slightly paraphrased here):

(1) Form a multifunctional team.
(2) State the project's mission and objectives.
(3) Select a team leader by consensus.
(4) Establish the team's ground rules for operations.
(5) Choose a name for the team.
(6) Prepare a computer-driven critical-path chart that lists the project's tasks and milestones.
(7) Determine risks.
(8) Estimate the project's cost.
(9) Establish preliminary specifications for the product.
(10) Prepare for a senior management presentation.

Many of these steps or phases would apply equally well to projects being set up for implementation by one person directing a group.

Four steps in project writing that will be discussed in detail in the remainder of this chapter are: (1) Determining the technical feasibility of a proposed project, (2) Setting priorities, (3) Budgeting, and (4) Establishing a schedule. The reader will understand that the project includes development of both the container and the food placed in it, i.e. the total package. Determining the marketing potential (including sales volume at each checkpoint) and marketing strategy, profitability, and distribution methods, and defining regulatory problems are points that will be treated in much less detail in this chapter, though consideration will be given to them later in the book.

EVALUATING THE TECHNICAL FEASIBILITY OF A PROJECT

The level of difficulty of a proposed project must always be an important consideration in deciding whether or not the project should be approved. There are many wonderful products that can be imagined but which have very little chance of being brought to fruition, given our present knowledge of science, technology, and engineering. It is the role of the technical participants in the project evaluation procedure to counsel management on the difficulties that will be involved when trying to balance the probable requirements of the project in terms of manpower, time, equipment, and funds with available resources. In the extreme cases — those products that are very simple and those that are well nigh impossible — the decision is easy, but in many intermediate cases a quantitative estimate of probability of success has a wide range of error, and management should be made aware of this.

Desrosier and Desrosier (1971) assigned levels of technological difficulty to four categories of new food products. They were principally concerned with judging the problems and costs involved in getting a new item manufactured, not so much with performing the laboratory experiments that are such an essential part of nearly every development project. The Desrosiers' four categories, in increasing order of difficulty, were as follows:

(1) The present factory, equipment, and technology can be used with a new formulation, processing sequence, etc.

(2) The existing factory can be used, but some minor new technology and equipment will have to be implemented – i.e., a new step in processing or packaging is necessary.

(3) The existing factory can be used, but new types of technology and equipment will be required – i.e., the processing method and/or the packaging equipment are different from those currently available.

(4) A new factory, new technology, and new equipment are required – i.e., raw materials, processing equipment, packaging, warehousing, and distribution systems all differ from those now being used.

The above arrangement formalizes the intuitive recognition that projects requiring the most complex and expensive changes in an existing operation will be the most difficult and costly to implement successfully.

SETTING PRIORITIES

"Priority" means (among other things) "precedence; superiority in rank, position, privilege, or the like." So far as lists of projects are concerned, the priorities assigned to projects describe the sequence in which they can call upon the product development resources of the company. In theory, giving priority #1 to a project indicates that workers spend their time on that project to the exclusion of others until (1) the project is completed, (2) the priority is changed, (3) data is being awaited (as from a shelf-life test) before the next step is taken, or (4) no one knows what to do next. In practice, it doesn't work that way.

The main purpose of the priority assignment is to allow R&D to assign its resources according to the needs of the company. Because the ranking of projects in order of their desirability is a decision affecting the whole company, the implication is that the CEO or his delegate will have the ultimate responsibility for assigning priorities, or, at least, for approving them. Until an idea has been fully described and has gone through at least a preliminary evaluation, it is impractical to set a realistic priority because there is not enough information on which management can base a meaningful judgement of the product's potential value to the company.

Factors to be considered in establishing priorities include, but are not limited to:

(1) Contribution the product can make to company's sales and profits both near-term and long-term.

(2) Possibility of successful completion of the development project within a reasonable time; "reasonable time" being defined according to the needs of the particular company.

(3) Investment requirements needed to get the new item produced and into

the market.

The author has found that the best method for identifying priorities is to affix a number to each project. Each number should be used only once, that is, there should be only one project with #1 priority, only one with #2 priority, etc. This is the most logical approach, but it is sometimes a difficult concept for management to understand; they frequently want to assign #1 priorities to virtually all the projects. The main problem with duplicated allocations of priority numbers (having several with #1 priority for example) is that it does not give the needed guidance to assignment of laboratory resources. In the absence of unambiguous priority assignments, lower level managers or bench workers will choose the project they want to work on (from the list of #1 priority projects, for example) while this should be a prerogative of top management. At the bench worker level, if choice is allowed, the project worked on is likely to be the one that seems most interesting at the time, fits closest the worker's specialty, is the most likely to be successful, etc., not necessarily the one that will benefit the company the most.

Priorities can be changed, of course, as additional data are accumulated or as management changes its goals. When a project has to be terminated, for whatever reason, the projects with lower priorities are moved up a notch, this step not requiring approval of top management.

It is not expected that everyone in R&D will work only on the project with #1 priority until that project is terminated. Work will be proceeding on a number of projects at any given time, since conditions may be such that work on, say, the #1 project could not be productive until further information becomes available. These situations occur repeatedly in the course of any development project. If the project list is of reasonable size, it is likely that someone will find time to do work on the project with lowest priority. If projects below a certain priority number are never worked on, it is a sign that the product development group has taken on more assignments than it can handle.

In the fortunate situation where the product development group is made up of several teams, each project may be assigned to a specific team, with each team having their own set of priority numbers. Thus, a team assigned projects #3, #8, and #16 would regard them as having priorities 1, 2, and 3, respectively, so far as their little world is concerned.

BUDGETING

Establishing a realistic budget and then making sure that expenses are held within that budget are essential elements of the procedure by which management can assure itself that the R&D department is operating in an efficient and predictable manner. The amount to be allocated to R&D, either in toto or by project, should be determined at conferences in which concerned parties relate a desired output to total company goals and resources.

The budgetary controls involved in R&D operations can be regarded as being of two basic kinds. Policy controls deal with the nature, scope and direction of the total effort, whereas operational controls deal with planning and evaluating specific projects during their life cycles.

Policy, or company strategy, will dictate how much money is to be allocated for overall R&D activity, including administration, depreciation, utilities, etc. This total will be divided into accounts for individual projects. Experience has shown that R&D efforts tend to diverge from the predicted path more often than most non-technical activities. Thus, a decision must be made whether to demand a strict balancing of the project goals against money spent, or to allow considerable freedom in fund allocation and disbursement as long as progress is being made in the general direction of company goals. There is no single answer that will apply to all firms, but experience indicates the desirability of promulgating moderately rigid guidelines, always with the proviso that technological breakthroughs may suddenly necessitate new funding requirements.

To be effective, project controls must alert the manager to any deviation from the plan – deviations in achievements, costs, or time. Budget variance analyses provide a comparison between the amount of money that has actually been expended and the amount that was originally allocated for a specific task. The department head must receive a monthly report of expenditures by account, if the the department's activities are to be planned in accordance with the budget. The financial reporting system is not adequate if it does not deliver such information promptly. The R&D manager should make it known he insists on receiving timely, accurate reports from the accounting department. He should verify all charges by matching them with vouchers, payrolls, orders, etc. Any unexplained items should be resolved by consultations with the comptroller.

There are few published guidelines on establishing budgets for R&D departments, especially for the food industry, and the ones that have appeared tend to be very general in nature (Fox 1975; Balderston 1976). It seems that allocations for R&D are often based either on a percentage of sales in the preceding year or on whatever competition is spending multiplied by some arbitrary factor. Management often feels uncomfortable about allocating funds to research because it differs from many other company activities in that there may be no significant achievements for long periods of time and there is great difficulty in assigning a monetary value to the discoveries that are made.

Continuity of the R&D effort over a period of years is necessary if it is to be fully effective. Research expenditures cannot be alterately expanded and contracted without destroying much of the department's efficiency. Shifting goals make for budget uncertainty, and major changes in management's objectives during the fiscal year should always be accompanied by a prompt assessment by R&D of the effect the changes might have on the department's expenditures. This study, which must be realistic and not opportunistic if management's confidence is to be retained, should be distributed to the heads of other departments which might be affected by changes in R&D funding (sales, production, and marketing, for example). It is probably better strategy to accept relatively minor changes in management direction without asking for increases in the R&D budget.

It is usually the case that more projects are recommended than can be adequately funded. Souder (1970) lists four kinds of techniques that have been used, or suggested, for selecting from the list of proposed projects those that should be supported; these are capital budgeting formulas, cost prediction formulas, scoring and ranking methods, and resource allocation. Each of these systems will be discussed briefly in the following paragraphs.

Capital budgeting systems are based on standard calculations of return on investment and discounted rate of return, modified by the inclusion of risk parameters such as the probability of commercial and technical success. A computed value for each project is discounted for the estimated risk. Projects are then ranked in order of their discounted values, and successively less promising projects are selected until the total allocation of funds has been exhausted. Uncertainties of this method result from the evolutionary nature of a development project, which causes constantly changing prospects of success or failure, as well as the effect of unexpected results from competitor's R&D activities, which may suddenly make the whole project obsolete. It is also quite apparent that raw estimates based on little or no reliable data often enter into the equation.

Cost prediction formulas are based on the premise that historical relations exist between the cost of a project and the sales of the resulting product. This is at best a vague relationship suitable for use only when more accurate projections are not available.

Scoring and ranking criteria are often used for selecting projects. In such systems numerical values are assigned to several factors, such as product life, patentability, market size, competition, and project cost and duration. The raw values or values weighted by some previously chosen factor are added together, and the projects then ranked in order of their total scores. The obvious difficulty in these methods is the uncertainty involved in deriving the assigned values.

Resource allocation techniques are intended to provide bases on which to decide how much to spend for a given project out of a series of projects, taking into consideration the total company situation now and as projected for the future. Data collected from all levels of the organization are submitted to computer analysis to determine the relative value of competing projects funded at various levels.

Swander (1992) stated his belief that top management can better understand R&D budgeting if the projects are classified into three categories: compliance, cost advantage, and differentiation. He defined these categories as follows:

(1) Compliance improves employee health and safety, ensures product

safety, or provides compatibility with environmental regulations.

(2) Cost advantage projects generating short- or near-term competitive advantages through changes in sourcing, formulation, processing, packaging, distribution, or sales.

(3) Differentiation creates long-term consumer value and competitive distinction based on flavor, variety, convenience, nutrition, or image.

There are several procedural details that experience has shown to be important. For example, budgeting should allow stability in staffing; large changes (up or down) in the number of technologists assigned to product development lead to inefficient utilization of manpower. Sudden influxes of new technologists require diversion of experienced personnel from project work to train, organize, and supervise the newcomers. It is usually found that the new workers require additional equipment for their projects, and this leads not only to unexpected expenses but to further delays in maximizing the employee's contribution to project advancement. On the other hand, terminations on a large scale not only create morale problems with the remaining technicians, they lead to inefficiencies in manning essential projects. For these reasons, sudden changes in the budget allocations for personnel should be avoided wherever possible.

The cost impact on other departments of new product projects is difficult to predict because it varies so much from project to project. If existing equipment is used in trial runs, the manufacturing department can frequently absorb most of the costs of preliminary trial runs with little out-of-pocket expense (though they may object to doing so). Very high costs may result if the process requires unique devices that have to be designed especially for the project. Packaging materials (design and printing) and consumer research often turn out to be the most expensive elements of new product development — at least up to the point of actual introduction, when advertising and promotional expenses swamp everything else. To some extent, these costs are under management control and depend on the impact to be made on the market.

Materials used for experimental runs in the plant should be charged to an R&D account, less any recovery for usable product. Provision should have been made in the budget for the cost of these tests. Often, it will be found that some or all of the experimental product can either be recycled or packed off as salable product, in which case R&D should be credited with their net value.

Indirect costs that are difficult to associate with specific projects include library expenses, telephone charges, secretarial and clerical salaries, administration allocations, utilities, and building depreciation. These charges are often allocated to projects in the proportion that the project budget bears to the department budget. If project accounts are to be charged for occupancy or similar overhead items, the exact amounts of these charges should be agreed upon at the beginning of the budgeting period.

A few science administrators have been extraordinarily successful in creating a kind of mystique that has led to such things as the publication of numerous articles recommending a "hands off" style of management and a science-centered administration as the ideal environment for "creative" R&D. Tolerance of top management and stockholders for such attitudes has caused some research directors to reach the mistaken conclusion that attempting to measure of the contribution of research to their company's financial welfare is somehow unethical or sinful.

In the long run, this is poor R&D management strategy. When hard times come and cost cutting is required in order for the company to survive, as it was for many companies in 1992 and 1993, R&D groups relying on such philosophies suffered much more than those relying on a more realistic appraisal of their reasons for existence. Very few companies can afford to support a group of dilettantes, scientific or otherwise, in the current economic climate. Furthermore, this kind of mystique has been largely pre-empted by marketing.

Product development groups have been in a somewhat better position than pure research sections. Personnel concerned with development have always been expected to produce results because the effects are highly visible and relatively easy to quantify. Development projects are funded with the idea that a payoff will occur within some finite time period. Even here, however, much stricter budgetary controls can be expected in the future with more accurate attribution of development costs to specific products. However unpleasant and unscientific it may seem, the science administrator is going to find that a knowledge of budgeting strategy is essential to the proper performance of his job.

RECORDING PROJECT INITIATION

There are undoubtedly dozens of more-or-less satisfactory methods of recording the progress of a project from its start until its completion, each of them having specific benefits for a typical kind of company. In Figure 4.1 is shown a suggested form for securing approval of a project in a small company. By combining all the needed descriptive information and approvals on one sheet (with attachments, as necessary) the recording and filing tasks will be greatly simplified. It is to be expected that each company will want to modify this form to meet its particular strengths and limitations. Large companies will necessarily require more extensive documentation, but the sample format shown here can still be useful as a basic guide for designing the multiple forms that may be needed.

Generally, the circulation, collection, and storage of these forms is best assigned to the director of new product development; in most small companies this will very likely be the head of R&D.

TO: Chief Executive Officer Date:______________

This proposed project is a [new/revised] project. The project it will replace, if any, is ____________ . It is based on Idea #____, which can be described briefly as ______________________________

The following steps have been taken to evaluate the feasibility of the idea:____

____

The following officers of this company have reviewed the idea and have approved further action, with reservations as noted:

1. Marketing/Sales
2. Production/Engineering
3. Legal
4. QA/QC

The [product/process], as currently visualized, is expected to meet the following specifications when it is fully developed:______________

____

The project will not require additional personnel, except as follows:

____

The project will not cause delays in projects already being worked on, except as follows:_______________________________________

____

Costs and progress are at present estimated as follows:

End of	Status of Project	Total costs
Month 1	______________________	________
Month 2	______________________	________
Month 3	______________________	________
2nd Quarter	______________________	________
3rd Quarter	______________________	________
First year	______________________	________

If this project is approved, R&D recommends that it be assigned priority number _____ in their list of active projects.

We approve the above recommended development work:

Signed and dated:

Marketing/Sales ______________________

Production/Engineering ______________________

Legal/Regulatory ______________________

Quality Assurance ______________________

Based upon the above information, I agree that work should be started on this project:

Signed______________________ Date:______________

Chief Executive Officer

Figure 4.1. Project proposal form for a small company.

BIBLIOGRAPHY

BAIRD, B. F. 1983. The Technical Manager; How to Manage People and Make Decisions. Lifetime Learning Publications, Belmont, CA

BALDERSTON, J. 1976. The research budget. Master or servant? Res. Dev. *27*, No. 8, 52, 54.

BASS, L. W. 1967. The planning of innovation. Chem. Ind. *1967*, 1671-1675.

BUSS, D. 1993. Speeding up the food industry. Food Business *6*, No. 7, 10-11. 13.

CLARK, P. J. 1987. Food manufacturing: Opportunities for improvement. Food Technology *41*, No. 12, 56, 57-58.

CLAYTON, J. T. 1987. Flexible manufacturing systems for the food industry. Food Technology *41*, No. 12, 66-67, 70.

DESROSIER, N. W. and DESROSIER, J. N. 1971. Economics of New Food Product Development. AVI Publishing Co., Westport, CT

FOX, H. W. 1975. Three strategies for R&D budget authorization. Food Prod. Dev. *9*, No. 1, 74, 76.

HIMMELFARB, P. A. 1993. New product development. Today's Chemist at Work *2*, No. 2, 56-57.

JAIN, R. K., and TRIANDIS, H. C. 1991. Management of R&D Organizations. John Wiley & Sons, Somerset, NJ

MARTENS, A. E. 1984. Make sure your R&D is as good as it can be. Research & Development *35*, No. 11, 86-88.

O'BRIEN, D. R. 1989. Managing for optimization of technical creativity. Food Technology *43*, No. 2, 82, 84-87.

SHIPMAN, F. P. 1978. Product development procedures. Cereal Foods World *23*, No. 3, 131-133.

SOUDER, W. E. 1984. Project Selection and Economic Appraisal. Van Nostrand Reinhold, NYC

SWANDER, D. C. 1992. Creating winning budgets: Improving your most important product. Food Prod. Design *2*, No. 3, 59-61, 64.

TWISS, B. C. 1986. Managing Technological Innovation, Third Ed. Longmans Group Ltd., Harlow, Essex, England

WHITE, V. P. 1987. Handbook of Research Laboratory Management. ISI Press, Philadelphia, PA

chapter five

PROJECT MANAGEMENT

INTRODUCTION

After an idea has been approved by management as being suitable for further development, R&D has the task of bringing the concept to fruition by conducting experiments and engaging in other activities appropriate to their department. To obtain a timely and efficient resolution of the many problems involved in developing a product, it is necessary that the idea be transformed into a well-defined project and that the project be administered in a tightly controlled manner. Most of the responsibility for achieving these results falls on the product development group, which may be called product development, R&D, new ventures group, or something else. Whatever it is called, it must have sufficient responsibility, funds, and authority to enable it to accomplish its goals, and it must be managed efficiently.

PROJECT CONTROL AND REPORTING

When a project has been established, with clear-cut guidelines for the new product's characteristics and a budget approved for its implementation, a series of events is initiated that must be carefully planned and efficiently managed to maximize the probability of technical and marketing success. Many systems and techniques for achieving these goals have been suggested and many different procedures are being used with varying degrees of success. It is probably fair to say that no plan can be devised that would be satisfactory for all companies at all times, but some general principles can probably be accepted and implemented by most food companies. Outlines of plans for controlling new product development step by step will be presented in this section. They can be used as a basis for designing product development procedures suited for individual development groups.

These plans are based mostly on project control systems instituted by the author at several food plants where he directed R&D, with modifications made to remedy some of the shortcomings observed in actual practice. The first plan is suitable for companies of medium to large size. It requires the participation of several managers on an ongoing basis. The second system is suitable for plants that are smaller but still have a well-defined group dedicated to product development; it is less detailed and more flexible than the preceding plan. Each plan has different advantages and disadvantages, and their value for a specific company depends on company size, degree of commitment to new products, and the talents of the product development staff, among other factors. Finally, a third product development system suitable for small plants will be described. This method of project control is suitable for companies having only one or two individuals working on product development.

A Project Control System for Large Companies

The key operating element in this plan is a Product Planning Committee (PPC) composed of the president of the company, the director of research and development, the head of marketing, the chief finance officer, the manager of production, and such other executives as are thought to have an immediate interest in, and some responsibility for, new product development. The committee appoints as administrative coordinator a secretary who issues agendas, takes notes of the transactions, issues minutes, and maintains records of the individual projects.

In this plan, entry of an idea into the system occurs as a result of the submittal of a new product suggestion to the secretary of the PPC. A standard form is made available to all employees for this purpose. Normally, the originator of an idea will fill out the form himself, but in some cases the secretary will make these initial entries. The secretary assigns a number to each entry received and keeps a record of the number, date of receipt, and a brief description or title of the idea in a journal maintained for this purpose.

The secretary's function in this respect is strictly clerical, i.e., he performs no screening except in obvious cases of irrationality or incomprehensibility. If necessary, additional information is solicited from the submitter in order to clarify vague or inadequate terminology. The proposal, with such additional information as he has been able to acquire, is either read by the secretary to the PPC at its next meeting or distributed in written form to the members in advance of the meeting. The PPC will either reject the proposal outright, ask for clarification from the submitter, or refer the idea to an idea analysis team.

The function of the analysis team is to make a study of the marketing and technical aspects of the idea. The study will be sufficiently extensive in scope to allow the PPC to determine if (1) a marketing opportunity exists for the suggested product, (2) the company is in a position to take advantage of the marketing opportunity, (3) development of the product is feasible from a technical standpoint, and (4) capital costs of getting the product into distribution can be afforded. The data at this stage will necessarily be crude and incomplete, and embody a considerable degree of uncertainty.

Experience has shown that most of the suggestions entering the system either will be close matches to others that have been considered earlier, will not conform to the the company's objectives, or will be clearly nonfeasible for technical reasons. If the committee decides the proposal has appreciable merit, a follow-up will be made by a representative (e.g., the secretary) who is assigned to extract from the submitter and other sources as much information as is necessary to allow a reasonably accurate evaluation of the proposal. A written report containing the findings is then submitted to the PPC. After considering this report, the committee either abandons the idea or assigns an analysis team consisting (for example) of one person from marketing and one representative from R&D.

After the analysis team has completed its study, a brief but comprehensive report is submitted to each member of the PPC. At the next meeting, the committee decides to (1) request additional work, (2) abandon the idea, or (3) advance the proposed product to the level of a priority project.

If it is decided to establish a priority project, the analysis team is disbanded (at least, so far as this idea is concerned), the project is given a name and assigned a place on the priority list (all priorities are temporary and can be changed at any meeting of the PPC), and a product team is assigned. Unless there are very cogent reasons for a different composition, the product team will include representatives from marketing, R&D, and production. Engineering may be included in place of production or in addition to production, if it is a separate department. It is not necessary to have permanent representatives from accounting, sales, quality control, or the legal department, but each of these groups will be called upon to provide input as the need arises.

The product teams are delegates of the PPC and thus of the president, and operate with his authority to gain action on necessary development problems and to coordinate the activities of the several departments. Team members will not ordinarily be department heads. If their efforts are successful, their final act will be the submitting to the PPC of a "Proposal to Manufacture and Market", a document that will recommend authorization of the funds required to produce and sell the new product. More will be said about this document later in this discussion. Alternatively, the team may at any time recommend that the project be abandoned because of technical nonfeasibility or adverse marketing information.

In the interim of months or years between the assignment of a product team and their issuance of a Proposal to Manufacture and Market (PMM), the necessary research, development, engineering, financial, and marketing studies will have been completed. Occasional reports (not necessarily on any fixed schedule) are made to the PPC, and informal minutes are kept of the team meetings. Each project is assigned an account number to which expenditures for time, materials, and services must be charged.

Each member of the team is responsible for obtaining the successful participation of his department in the development work necessary to produce the new item. If his own influence is not enough to obtain the needed cooperation from his departmental colleages, a member can enlist the aid and authority of his department head. Problems in securing cooperation of other departments, plans for future work, estimates of phase completion dates, funding requirements, and chances for success are discussed at team meetings.

From time to time, the composition of the team may be changed by the PPC as the need for participation of different departments changes. Prototypes of the new product (including package designs) plus market research data are submitted to the PPC as they become available. Based on these exhibits and the reports of the product teams (presented either orally or in written form), the PPC decides at each meeting either to abandon the project, continue it as planned, or alter its direction. In the normal course of events when several projects are under continual consideration, at each meeting some of the teams will present additional data of importance and the committee will decide to continue the project. If work is to continue, a priority is assigned or continued.

During the course of the development process, production line designs will be completed, product costs will be finalized, ingredient sources will be determined, quality assurance procedures will be agreed upon, legal clearance will be obtained, and all other arrangements will be brought to the point where successful production can be ensured. The results of these studies will be set forth in the PPM.

As stated previously, the PMM is essentially a proposal to top management that expenditures be authorized for the manufacture and marketing of a new product. An important feature of the plan is a complete specification of the new product. This will give firm, unequivocal requirements for each important characteristic of the product and of the manufacturing process required to achieve those characteristics. None of these requirements is to be changed without the prior concurrence of all interested parties, and production is charged with the responsibility of seeing that the quality assurance provision are met.

In case of disputes over the provisions of the plan, e.g., when production expresses doubt about its ability to consistently meet the quality standards deemed necessary by marketing, arbitration is performed by the PPC. Acceptance of the details of the plan by each responsible department head results in the transformation of the PMM into specifications that describe the commercial product and the processes used in its manufacture.

Project Control System for Medium-size Firms.

The second plan for controlling development projects for new products will be found suitable for many companies, including those of intermediate size. In this system, consideration of a new product is initiated by the submittal to the president of the company (or his representative) of a proposed new item. The suggestion will contain, as a minimum, a brief description of the new item together with a statement of its proposed advantages over other available foods of the same type. The proposal can originate from any of the division heads within the company, from an employee suggestion plan, or from a source outside the company.

Unless the president finds the suggestion to have no merit, he will forward the proposal to the head of marketing, together with a directive to ascertain the marketing potential of the proposed item. If marketing concludes that the proposed new product has inadequate sales potential, its recommendation for abandonment of the project is sent to the president for final disposition. This recommendation will be accompanied by a detailed justification, including pertinent data, for the adverse decision. The president will base his further actions on marketing's memorandum plus other advice and information.

If Marketing should, on the basis of its tests, experience, and surveys, deem the proposed new product to have a reasonable chance of success, it establishes certain broad and tentative specifications of a preliminary nature. These will include:

(1) Maximum acceptable unit sales price and minimum permissible profit margins.

(2) Product (package) size.

(3) Minimum shelf life at some definite storage temperature (e.g., 0°F, 40°F, or 70°F) with criteria specified for determining the end point.

(4) Production capacity to be available for start-up and one year thereafter.

(5) General characteristics of the product in terms of appearance, flavor, and texture.

(6) Limitations of home preparation requirements in terms of equipment needed and time expended.

The container and label are to be considered as part of the final product and each of the steps requires the food material, the container, and the label to be at some specific stage, though not all components will necessarily have achieved the same percentage of completion.

These preliminary specifications, for which marketing assumes the responsibility (though they must be assembled with the assistance of other departments), are incorporated into a memorandum. The original is forwarded to R&D and is accompanied by both the initial proposal and the directive received from the president. A copy of the memorandum is forwarded to the president. After an analysis of marketing's recommendation, the R&D staff either (1) submits to marketing a proposed schedule in which an estimate of the probability of the project is stated as a percentage, giving also the time required for completion and an estimate of total man-hours and costs required for the technical work, or (2) indicates to marketing that the product as specified is nonfeasible due either to technological problems or to some unrealistic limitations established by marketing. In the latter case, R&D will make suggestions for removing the obstacles by modification of marketing's requirements, if that is possible. Then, marketing and R&D will confer in an attempt to reconcile their differences of opinion. If agreement is not reached in this conference, the file on the matter is submitted to the office of the president for final action.

In the case of a favorable report, marketing will (after having studied the plan of R&D) submit a memorandum to the president in which it is recommended either (1) that the schedule of time and costs for development be accepted, or (2) that consideration of the project be discontinued because of excessive research costs and/or the limited probability of success. In either case, written concurrence of the president is required before the project can be considered terminated.

If the president agrees with marketing that R&D should perform the work it has outlined, a formal research project is initiated and appropriate funds to support it are allocated. Simultaneously, a committee consisting of one representative each from marketing, R&D, and production is set up to coordinate and expedite all activities needed for getting the product on the market. This committee has primary responsibility for the achievement of successful production. Members of the committee are appointed by the various divisions, with due notice to and concurrence by the president, and will ordinarily include persons expected to have close and continuing contact with the actual work being performed. For example, R&D might appoint as its representative the technologist responsible for doing the experimental work on the project.

Each member of the committee is responsible for coordinating the activities of his department with the work being done by the other departments. Each member receives copies of R&D progress reports on the project, and the representatives of marketing and production are responsible for informing the other members of the committee of all pertinent actions taken by their respective departments. This information must be relayed to the other members within the shortest practicable time after the action is taken; the same day, if possible.

Written concurrence of all committee members must be obtained before major actions, such as the purchase of production equipment, are undertaken. Information can be exchanged and concurrence obtained during committee meetings if desired and, in this case, the transactions are entered in the minutes of the meeting.

All decisions made in committee meetings are set down in the minutes, copies of which are circulated to all interested parties, including the president. The original of the minutes is retained in the file maintained by R&D. R&D's representative on the committee acts as the secretary and is responsible for transcribing and reproducing the records of the meeting and for circulating them to the other members as soon as possible after each meeting. If either of the other committee members objects to the accuracy or completeness of the minutes or wishes to make additions or corrections for other reasons, he places in the file a memorandum setting forth his opinions. Copies of this memorandum are sent to the president.

The purpose and final aim of the committee is the production and marketing of the new product, and its work does not stop prior to that stage unless the project is terminated by authority of the president of the company. When the product is being produced in a manner satisfactory to all concerned, the committee will disband.

In this type of system, the final goal of the committee is an operating production line and distribution system rather than the plan described for the preceding system. Fewer people are involved at any one time and the paper work is reduced as compared to preceding plan. As a consequence, the progress of work is not as well monitored and the chances for slippage, misunderstandings, and failure to perform are increased, but, on the other hand, personnel interactions are more direct.

Project Control System for Small Companies.

For companies having one to about five persons working full-time in R&D or product development, or perhaps one person who can devote only part time to the function, the formal schemes for project control described in the preceding sections are too elaborate. What is needed for these companies is a plan insuring collaboration between R&D, marketing, and production in an manner that promotes timely completion of assigned projects. This cooperation and interchange of information can be achieved only by frequent and regularly scheduled meetings accompanied by the circulation of written and detailed reports to those persons who must perform some of the activities but who cannot, for one reason or another, attend the meetings.

The CEO should assign a product development committee consisting of, as a minimum, the head of marketing, the head of production, and the head of R&D. The CEO should also participate in the meetings, if it is possible to get him to do so. The head of accounting and a legal adviser should be invited to the meetings, as needed. Meetings should be scheduled no less than two weeks apart. Weekly meetings are preferred, with unscheduled meetings being held when some new decision point is reached. The purpose of these meetings is to inform the participants of the progress made by each of the members and to obtain a consensus on future actions. Decisions to continue the project, change direction, or terminate the project are made in these meetings. Each member accepts responsibility for performing certain actions which are necessary to further the project.

Whatever the frequency of the regular meetings, they should always be held on a given day at a given time, so members will have less opportunity to forget them.

A series of printed forms is suggested for recording decisions, reporting results, and assigning responsibility for needed action. As a minimum, the forms will be (1) project initiation form, (2) project status form, (3) report of committee meeting with assignment section, (4) cost of product calculation, (5) legal/compliance opinion, (6) product and container specifications, (7) report of the Purchasing Department, and (8) plan for introducing and marketing the product. If the CEO does not apply his full authority to motivating the committee and supporting its actions, it is useless to expect rapid progress.

PROJECT STATUS REPORT

1. Title of the Project_________________________________
   Report covers period from ______________ to ______________
   Description of the desired [product/process]:______________

2. Summary of status at beginning of reporting period:____________

3. Accomplishments during the reporting period: ______________

4. Present status of the project:________________________

5. What remains to be done: ________________________

Plans for the Next Reporting Period

R&D Objectives: ____________________________

Production/Engineering Objectives:____________________

Marketing/Sales Objectives:________________________

Production Objectives:__________________________

Legal/Regulatory Objectives: ______________________

Are efforts needed by other departments or outside agencies?

Estimated expenditures and commitments to date: ____________
Projected expenditures:
   (a) Next reporting period ______________
   (b) Total project ________________
Estimated completion date (satisfactory product coming off the line):

Sign offs, with dates:
   R&D:
   Marketing:
   Production
   QA/QC:

Figure 5.1. Project status report form for a small company.

It is unusual to find in any company three supervisor types who are willing to submerge their narrow and immediate interests to achieve a distant goal, especially when achieving that goal may not add any particular luster to their reputations. The CEO must provide the needed impetus.

Of the several important forms or records that will be important in the product development efforts of a small company is the Project Status report, a suggested example of which is shown in Figure 5.1. Its purpose is to keep everyone connected with the project informed as to its status and the plans for future work. The illustrated document has been formatted so as to fit on one sheet but, in practice, it is almost certain that attachments will be required to fully explain the achievements and scheduling.

Other Examples of Product Development Systems

Actual examples of product development procedures are rather infrequently described in published articles, presumably because of security considerations, but a few publications can be found dealing with procedures in effect, or which have been in effect, at some of the larger consumer products companies. The following examples are not necessarily restricted to food product development.

Kill (1965) described a system that can be assumed to resemble the sequence of events involved in the conducting of a product development project at Nabisco. These are the major phases:

1. The idea is submitted on a form addressed to the R&D laboratory. This report includes, besides administrative and control data, a project title or description, the ultimate objectives of the project, potential company benefits, and estimated cost of implementation.
2. Industrial intelligence information, such as availability of competitive items, potential demand, etc., is collected.
3. A profit study with projected sales volume is made.
4. The research department calculates cost of manpower, materials, and equipment needed to develop a prototype acceptable to the sales and merchandising departments.
5. The production manager lists requirements for his department as he sees them.
6. The product development committee considers a wide variety of aspects of the contemplated product as it affects the company's future before it passes on the desirability of the project. The committee may be composed of some or all of the following: members of the Board of Directors, owners, general executives, the president, the vice president, the secretary, the treasurer, and the comptroller.
7. If approved, the project is assigned a priority and given a time limit.
8. Assignment of a project to a product development group is followed by the production of prototypes. Progress is communicated to and discussed with the production manager, who can commence preliminary design work

or express opinions as to the feasibility of the process. Liaison with marketing is maintained in a similar manner. Sometimes these liaison functions are performed by the product manager, who is usually a member of the marketing department. Packaging development will be going on at the same time.

9. A detailed description of all production costs is given to the finance department for a profitability study. If, as is often the case, adequate profits cannot be projected using the costs and prices assumed at this stage, the various departments are asked to modify their requirements and adjust their calculations.

Hormel uses a client-for hire system by which research costs are charged to marketing for specific projects. New product ideas are first evaluated by marketing. Those deemed desirable are discussed with the director of R&D to ascertain their technical feasibility before a project is set up. If it is decided to proceed, the following sequential steps are taken in what can be called Phase I.

1. The objective is defined. What is the project expected to accomplish?
2. A justification is provided. How will the company benefit?
3. Preliminary specifications are written. The product or process is described as accurately as possible, including the chemical, physical, and organoleptic properties, package design, cost limitations, etc.
4. Priority, timing, work required, and other administrative topics of the project are established.
5. Supporting data are supplied to facilitate the determination of scientific feasibility and the design of experiments.

When the above steps have been completed by Hormel personnel, a project leader is selected. He searches the literature, decides on an approach, outlines the method of attack, and estimates requirements of time, money, and skills. Marketing decides whether or not to continue on the basis of this information. This is Phase II.

The project is readied for final review in Phase III. After presentations of their viewpoints by all responsible functions, the project receives final approval or rejection from a group vice-president, the R&D director, and either the corporate vice president or a divisional manager.

In Phase IV, R&D personnel take over responsibility for the project, and a team is selected to work for the project leader. The team outlines the experimental approach to be used, and issues weekly reports covering money and time expended. Monthly reports are made describing technical progress.

The Hormel system seems to me to be logical and to have a number of other good points. It would repay detailed study by anyone intending to set up a new product development system or to revamp an existing one.

According to Alleman (1967), these were the key points in Kroger's bakery product development program:

1. Have a definite goal or objective for each project.

2. Follow a set plan.
3. Keep eyes and ears open for new ideas.
4. Be flexible and adjust to changing times.
5. Be aggressive. Don't be afraid to try new ideas.
6. See that all work is done carefully and accurately.
7. Be sure you know the cost of new products.
8. Get the best possible consumer evaluations before going too far.
9. Accept help from any source. Don't be too proud.
10. Make every project justify itself. Don't do useless work.
11. Have enthusiasm for all new developments. If you are not impressed with your own work, don't expect enthusiasm from others.
12. Have a positive attitude at all times and drive the projects to the goal of goals – more profits.

To be sure, most of the above points are very general and not too helpful in establishing the details of new product development system, but they have a certain inspirational or philosophical value.

S. C. Johnson Co. has, or had, sponsor groups for new product development. These groups include (1) the man who had the idea, (2) the R&D man who is going to supervise the lab work, (3) the man who has to sell the product when it is completed, (4) financial and production personnel, and (5) a member of the new products department. Details of procedures were not given in this publication.

Wolf (1959) described the product development system established at Arnold Bakeries. The essential steps are:

1. Ideas are obtained from employees, executives, and outside sources.
2. Market Reserach evaluates the idea on the basis of concept tests with consumer panels.
3. Management selects those ideas which it can manufacture profitably, and assigns priorities.
4. Development work is started.
5. Final product characteristics are specified by management and market research after the initial prototypes have been examined.
6. Experimental plans are drawn up by the laboratory.
7. Test bakes are made and evaluated by R&D personnel on the basis of the specifications established by management.
8. Prototypes that appear to meet the requirements are evaluated for acceptability by the market research department using consumer panels.
9. Acceptable items are coordinated with production to develop suitable processing methods. Time studies enable labor costs to be predicted.

The management of product development in a large, diversified international company with substantial food interests was described by Cotton (1970). Requirements in different countries vary so much that a single research, development, and engineering facility for the entire company was not considered feasible. Therefore, decentralized profit centers and laboratories were set up in marketing regions unified by common requirements, while a world headquarters technical group takes responsibility for overall policy guidance, information transfer, and budgeting. Maximum self-sufficiency in the units is encouraged. The projects they work on are developed by the national research department in close cooperation with top marketing and production executives. The World Headquarters Technical Department receives copies of project reports and maintains liaison with the national laboratories at appropriate management levels.

In setting up a project, according to Cotton's plan, the first step taken by the national company is to collaborate with marketing personnel in defining objectives. The second step is setting up a three-man team generally composed of representatives from production, marketing, and R&D. These development teams can cut across normal lines of command to achieve their objectives. Each week a progress report is made to top management, and assistance in obtaining action is requested, if necessary. Research supervisors meet every three months with the director of research and the manager of research administration to compare results with planned schedules. The national companies also have research committees consisting of the chairman of the board, the president, and the vice president of reserch that review programs and progress every month.

The project control system used by Sunbeam (Anon. 1976) consists of 21 steps which take a product from the concept stage through initial production. During the development process, all departments with an interest in the project have access to the research and engineering personnel who have been given primary responsibility for the product. This system is said to promote early discovery of problems and make solutions easier. Steps 1 through 5 involve drawing up a presentation to management describing the need for a product and the initial approach to be used. After a preliminary design is completed and tested in Steps 6 through 8, a report is presented to marketing, manufacturing, and general management. Final specifications lead to improved samples, which are demonstrated to marketing in Steps 9 through 12. Management makes a final review (Steps 13 and 14), and design details are transferred to manufacturing engineering, which finalizes production details with input from suppliers, distribution, etc., in Steps 15 through 19. After initial production begins, a review of the project (including cost and time goal achievements) is made (Steps 20 and 21).

At Green Giant, there are four stages in the new product development system (Ziemba 1966). These can be summarized as:

Stage 1. Establishing the new product concept. An idea is screened by the Director of Research, a Research Committee, and the Director of Product Development. If the concept is approved, a Project Planning and Status Record is started.

Stage 2. Making a technical and economic survey. The economic feasibility (profitability) of the product is thoroughly investigated by several departments of the company working with preliminary formulations and processing and packaging specifications prepared by the Director of Product

Development.

Stage 3. R&D Activity. If the Operating Committee approves further work, formulas are developed, sales projections and plans for promotions are drawn up, procedures and facilities are made ready, and all other important aspects of the product are studied simultaneously under the aegis of the Operating Committee.

Stage 4. Transition to Commercial Operation. All divisions share the responsibility of expanding operations to include the new product.

MEASURING COMPLIANCE WITH BUDGETARY GUIDELINES

As far as actual expenditures for product development are concerned, not much information on the practices of food manufacturing firms can be found in the published literature and there is some question as to how meaningful the figures would be anyway, since different companies would probably include different accounts in this classification.

The budgetary controls involved in R&D operations can be regarded as being of two basic kinds. Policy controls deal with the nature, scope and direction of the total effort, whereas operational controls deal with planning and evaluating specific projects during their life cycles. Policy, or company strategy, will dictate how much money is to be allowed for overall R&D activity, including administration, depreciation, utilities, etc. Out of this total will be allocated amounts for individual projects.

It seems that R&D efforts tend to diverge from the predicted path more often than non-research activities. Thus, top management must make a decision whether to demand a strict balancing of the project goals against money spent, or to allow considerable freedom in fund disbursement as long as progress is being made in the general direction of company goals. There is no single answer that will apply to all firms, but experience indicates the desirability of moderately rigid guidelines, always with the proviso that technological breakthroughs or other sudden and unexpected information may suddenly mandate new funding requirements.

To be effective, project controls must alert the manager to any deviation from the plan – in achievements, costs, or time. Budget variance provides a comparison between the work that has actually been accomplished and the amount that was planned. A monthly report of expenditures by account is essential to the planning of the department's activities. A budgeting system is inadequate if it does not deliver such reports promptly. The R&D manager should insist on receiving timely, accurate reports from the accounting department. He should verify all charges by matching them with vouchers, payrolls, orders, etc. Any unexplained items should be resolved in consultations with the comptroller.

BIBLIOGRAPHY

ALLEMAN, H. J. 1967. The practical development of new products. Proc. Am. Soc. Bakery Engineers *1967*, 287-292.

ANON. 1976. From concept to product — in 21 steps. Plastics Eng. *32*, No. 7, 1-4.

COTTON, R. H. 1970. Management of product development in a large diversified corporation. Food Technology *24*, No. 4, 36-38.

KILL, J. F. 1965. New product development in cakes, sweet yeast raised products, and cookies. Proc. Am. Soc. Bakery Engineers *1965*, 244-248.

SWANDER, D. C. 1992. Creating winning budgets: Improving your most important product. Food Prod. Design *2*, No. 3, 59-61, 64.

WOLF, A. 1959. From an idea to a finished product. Baking Ind. *111*, No. 1407, 23-28.

ZIEMBA, J. W. 1966. R&D guides Green Giant. Food Eng. *37*, No. 1, 98-100.

chapter six

PERFORMING TECHNICAL DEVELOPMENT WORK

INTRODUCTION

In this chapter, we will consider some of the procedural details involved in planning and performing the technical work that makes up such a large part of new product development. As a starting point, the administrative relationship of the technical development group with other departments in the company will be examined, and recommendations made as to the preferred arrangement. The organization of the technical department that is best suited for efficient and productive work will be discussed and analyzed. Then, methods that have been used for designing experiments, including those systems depending heavily upon computer assistance, will be covered. The importance of proper record keeping is stressed, and a procedure that has proved its value in practice will be given. The critical subject of sensory testing is the topic that follows, and lessons learned from long and painful experience with poorly designed and improperly conducted tests are presented. Storage stability is often given only brief attention in product specifications, sometimes with disastrous consequences – justifying a lengthy section on testing and modifying shelf life. Operations research, evolutionary operations, new venture groups, use of consultants, and other ways of systematizing and conducting experimental programs are discussed in a series of articles. The final section deals with timely termination of both successful and unsuccessful projects, a subject of considerable importance in maximizing the most effective use of R&D resources.

ORGANIZATIONAL STRUCTURES

The position of the new product development group within the company's table of organization has an important effect on the freedom of the group to design, conduct, and analyze the results of the experimentation necessary to quickly and inexpensively design a new food and its container. It is also a question that can cause bitter arguments within the ranks of top management, as executives in other departments maneuver for control of a group that has a certain prestige and enough productivity to cover some of their own lapses. Some – certainly not all – of the ways that have been used, and could be used to integrate the development group into the larger organization will be covered under the present heading.

Position of NPD Group in the Table of Organization

In many food companies, product development functions fall under the direction of a department of broader scope called Research and Development, or, sometimes, Technical Services. The former title implies the performance of some basic investigations, whereas a technical services group will often include the quality assurance function as well as applications services performed for industrial customers.

D. M. Schoengrund (1983), at the time Director-Corporate Product Research of the Hershey Foods Co., gave an example of corporate organization in which [basic] Research and Product Research were combined in a Department of Science and Technology. The Director of this department reported to "Senior Management," the latter group reporting to the Board of Directors.

We will consider in greater detail in the following discussion the more complex organizational arrangement, in which product development is performed by an R&D department, i.e., by a group of people specifically assigned to conduct all of the technical studies necessary in designing a new food and its container.

Because the success of the R&D department is essential to the health of the company and because its goals and methods of operation differ in many respects from those of the other departments in a manufacturing company, it should report directly to the president or CEO. In this way, its influence can be felt throughout the company and support at the highest level will be continually evident. The filtering of information through a nonscientific channel, such as would occur if R&D is given a more subordinate position, will very likely result in distortion of its views and failure to transmit important information. By making the director of R&D a member of the executive committee, the president and other members of that group can derive the utmost benefit from the advice and example of a scientist or technologist.

There must be a single administrator responsible for all R&D functions; fragmentation of the activities among other departments such as engineering or marketing is extremely inefficient and inappropriate. Process development, process engineering, and package design (with the exception of graphics), as well as product development, must be directed by one administrator so that the close coordination of activities which is so essential to success can be achieved.

The use of terms such as "research" and "product development" in other than their scientific connotation — for example, the use of "research" to denote consumer survey studies conducted by marketing, or "product development" to mean the sales program involved in the introduction of new or modified products — is inexact and may tend to confuse top management. Unless the R&D director is both forceful and persistent in making clear to the chief executive the difference between certain non-technical functions and scientific activities similarly named, confusion will inevitably develop as to the proper role of marketing and R&D in managing the company's new product activities.

It is usually desirable to restrict official contacts of the personnel of other departments with R&D personnel to formal channels of communication. Direct contacts at lower echelons can lead to dissemination of erroneous and sometimes harmful information. Many breaches of security in new product development can be traced directly to informal discussions in the laboratory. Also, free food has an overwhelming attraction for many people, and its reputed availability in the R&D area can lead to a stream of distracting visitors (and, sometimes, loss of valuable samples) unless access to the area is rigidly controlled.

Organization of the Department

In the Schengrund article previously cited, the Product Research Department included the sections New Product Planning, Product Development, Packaging, Engineering, and Quality Assurance. The only technical group not in this department was the organization devoted to basic research. There are several other ways of organizing and managing the product development function, some of them may be better, others worse, in terms of efficiency.

It is customary to separate the R&D department into a research section which deals with fundamental problems not directly connected with a specific marketable product, a development section responsible for product and process development, and an engineering section responsible for implementing on a commercial scale the laboratory or pilot plant processes of the development group. Lines of demarcation are not always clear and perhaps need not be. There should be no attempt to emphasize the differences between research, development, and engineering lest valuable contributions of unorthodox origin be ignored. In smaller companies, it may be best to eliminate administrative separation of these functions.

Line organization within the development staff can be formed in at least two distinct ways. Personnel can be divided into administrative groups either on the basis of the products or processes dealt with or on the basis of the disciplines involved in their work. By the former method, all persons working with vegetables would be supervised by one person, whereas those concerned with fruits would be in a different group. The "disciplines" method would place chemists in one section and bacteriologists in another, for example. There are refinements and variants of these systems which need not concern us at this time. Each method has its advantages and disadvantages, but, in general, the product-oriented scheme is better for relatively small companies while the discipline-directed system seems to work better for larger companies with a wide range of interests and a long-term development policy.

Miller (1986) describes possible project structures in essentially the same form, though more elaborately, perhaps. The remainder of this paragraph is paraphrased from Miller's discussion. The classical structure in *research* is based on the discipline, or field of science. Thus, for example, a research organization will have a chemistry group, a physics group, a life sciences group, and a material sciences group. Each of these will normally be led by a person trained in that discipline. In *development*, the more classical form of organization is programmatic. Thus, an organization is formed in the early conceptual period of a product or system, and grows with the need to develop the product. Development organizations thus tend to follow the product type.

Some companies may find that a completely different type of line organization is better for them. The system should, however, at least be consistent and it should be based on a conscious plan. Allowing groups to form and break up on the basis of short-term projects is poor policy.

An important requirement for any organizational scheme is that it must provide an environment that facilitates the performance of creative work. There are many ideas about how this can be accomplished, but proof as to the relative merits of the different schemes seems to be lacking.

According to Koning (1974), a good environment for research should include the following factors:

1. A scientist or group of scientists should be assigned to an actual and significant problem that (a) is related directly to a specific goal of the organization and (b) allows the scientist to develop (exercise) his discipline or specialty.
2. Within a problem area, scientists should be allowed the independence to develop studies necessary to solve the problem.
3. Scientists should be supervised by knowledgeable scientists and administrators who show continuing interest in the work, who attempt to fulfill the individual needs of the scientists, and who make a sincere effort to provide all the physical and technical support the scientists require.
4. Completely open communication should exist throughout the research organization.
5. Scientists should be given generous recognition for work well done, and, conversely, should be constructively criticized when work is poorly done.

Although the complete application of these tenets would doubtless be possible only in some highly idealistic organization staffed by completely unselfish personnel (a situation not known to be represented by an actual example in government, religion, education, or industry), the overall concept at least represents a series of desirable goals.

Setting up a rational system for controlling product development merely provides a framework or base for the day-to-day work that will, it is hoped, lead to a successful new product. Within the limitations of this book, it is not practical, or even possible, to describe all of the alternatives that should be considered by the research manager in choosing the methods he or she will use to achieve the desired objective. The purpose of the following paragraphs is to call attention to some of the procedural details that are particularly important and to point out alternative methods of conducting product development.

The initial step in any development project should be a thorough literature search to determine the existing state of the art. The search should encompass not only scientific and technical periodicals and reference books, but also patents, technical bulletins of ingredient and equipment suppliers, and the company files which may contain reports of previous work along similar lines.

Results of the initial literature search should be given in a formal report containing abstracts of the pertinent articles and a bibliography with all the papers reviewed listed in it. The report should be kept in the file permanently. This important step in the development project is frequently neglected, with the result that changes in personnel or other interruptions in continuity lead to the necessity for doing the work all over again. If time and funds are available, the literature search should be supplemented by discussions with workers in the field and visits to equipment installations approximating the type which would be used for the new product.

The techniques for designing experiments, conducting experiments, and statistically evaluating results are too complex to be covered fully in this chapter; a book, or perhaps a multivolume encyclopedia, would be required for a reasonably complete discussion of all aspects of these subjects. The best that can be done here is to skim the surface of some important points that may be of particular value to the reader. More thorough treatments can be found in the books by Walters (1965), Stanley and White (1965), and White (1988).

DESIGNING THE EXPERIMENTS

Much of the experimental planning used in food industry R&D is based to a great extent on intuitive approaches to condition optimization and product evaluation. This tends to give statisticians and theoretical scientists a bad case of the horrors when they are confronted with an explanation of the "reasons why" this or that was done and something else left undone. Nonetheless, substantial advances have been made and many succesful products have been developed by persons and groups operating by these less-than-rigorous procedures.

The lone experimenter working at the bench, as will be found in many small operations, does not have the time or materials to investigate all possible variables in finely graduated increments. He or she will, in many cases, change two or more variables at the same time, work without control samples, etc., substituting experience and skill for the greater amount of objective data that could be obtained if the experimental design were optimal. The purpose of the following discussion is to describe ways in which design of experiments can be improved to give greater assurance of the significance of results and reduce errors. It is not intended to dictate the implementation of rigorous experimental design immediately in all product development studies.

Seiler's (1965) summary of the steps in an R&D project can be paraphrased as follows:

1. Assemble data, study previous work of a similar nature, define the problem, and formulate a hypothesis.
2. Decide upon the design of the experiment and assemble the necessary materials, equipment, and personnel.
3. Conduct the experiments necessary to verify or disprove the hypothesis.
4. Arrive at a creative solution with regard to the relations between the variable being studied.
5. Verify the solution and extrapolate it to the full scale requirement.

According to Joglekar and May (1987), design of experiments is viewed today as a quality technology to achieve product excellence at the lowest possible cost. It is a method for optimizing product and process designs, accelerating the development cycle, reducing development costs, improving the transition of products from R&D to manufacturing, and facilitating troubleshooting.

MAINTAINING RECORDS OF EXPERIMENTS

In addition to the usual reasons for keeping notes on the experiments performed during the course of a development project, the necessity for patent purposes of documenting the date at which a discovery is reduced to practice dictates use of a formal recording system. A haphazard collection of loose data sheets is of little or no legal value for this purpose. The research notebook should be permanently bound, not looseleaf. Pages should be numbered and a record kept of the individual to whom the notebook is issued.

When the book is completed or its use discontinued for other reasons, a brief index of its contents should be made and filed. The index can be prepared by photocopying the table of contents prepared by the experimenter during his use of the book. The book itself should be stored in a lockable cabinet, a safe, or some other secure place when it is not being used.

Entries should be made in ink or indelible pencil. It is customary to draw a diagonal line across the part of the page left blank. The date the experiment was performed should be placed on each page. A witness who understands the substance of the data but who has not participated in the experiment should affix his signature (as a witness) and the date at which the page was signed. No marks, including corrections or comments, should be made on the page after the witness has signed it.

It has been recommended that the following items be included as part of the description of the experiment:

1. Reference number of the experiment, the subject or title of the project, and the date.
2. Object of the experiment – what is intended to be accomplished by it?
3. Apparatus and procedures – references can be made to published descriptions or procedures, or to those described on other pages or in other notebooks, but anything newly introduced should be fully described.
4. Starting materials – source and grade of ingredients; any special treatment that has been applied to them before the experiment; age and storage conditions.
5. Results, including:
    a. Products – details of all essential characteristics: appearance, measurements, disposition, etc.
    b. Processes – refer to engineering drawings or photographs when possible. Reference to research notebook number and page numbers should be permanently affixed to these drawings and photoprints.
    c. Samples – identify any samples retained or sent elsewhere for analysis. The number assigned to a sample could be a combination of the research book number, page number, and a sample designation mentioned on that page.

CONDUCTING ACCEPTABILITY TESTS

Publications on the theory, rationale, and history of taste testing have given us a rich literature resource which cannot be reviewed in detail in this brief article. A recent review of considerable value is that of Lawless and Claassen (1993). These authors suggest a sequence of acceptability tests that could be applied during new product development (based on a 1979 publication by Pangborn and somewhat further modified in the following list):

(1) Bench-top tasting. To familiarize technologists with product attributes and possibly allow the elimination of seriously defective formulas.

(2) Difference tests with lab panel. To determine whether or not samples can be differentiated.

(3) Directional difference tests. To explore sensory attributes responsible for observed differences.

(4) Scale attributes. The function being to specify intensity and duration with time-intensity techniques.

(5) Descriptive tests. To fully define important attributes such as effects of process or ingredient variation.

(6) Hedonic tests/pilot consumer panels. To establish degree of like and dislike.

(7) Central location or home tests with representative consumers. To test acceptability and probe consumer language.

For the home cook, the food testing panel is small and well-defined –

generally self, spouse, and kids – and the hedonic scale is short and emphatic – more, OK, and dump it. When a product development group aims for a national (or even an international) market, the input is likely to be a bit more complex. Herein lies the necessity and value of the art or science of designing, testing, and interpreting organoleptic tests for the guidance of food manufacturers.

Non-Product Factors Affecting Taste Test Results

As with many of the subjects discussed in this book, the body of information available in the literature and elsewhere is so great that any attempt to review all the available material in a work of this size would be futile. Instead, the author will discuss a few topics often neglected or treated superficially in the more comprehensive treatments that have been published.

Effects of temperature. It should not be necessary to emphasize the differences in acceptability ratings that accompany changes in the temperature at which the test material is consumed. Yet, there are seldom any efforts to maintain uniform temperatures in the test portions as they reach the hands of the subjects. The temperature effect on ratings may be slight, as in the testing of cookies or bread at room temperature, or substantial as in the cases of soup, pie, pasta, carbonated beverages, etc. Although the bulk supply may be kept in a thermos or other temperature control reservoir that maintains the test material within a relatively narrow range of temperature, the samples themselves may change substantially between the time the food or beverage leaves the bulk container and the time it reaches the panelist's mouth. Uncontrollable variations may occur due to a panelist's slowness in consuming the sample. Every reasonable effort should be made to maintain uniformity of sample parameters at the time of consumption, and in any case, temperatures should be taken of representative samples so that the technologist will know what is happening.

Influence of containers. An important but often ignored source of error in organoleptic testing is the contamination of storage and serving containers by odors and tastes. These contaminants include materials inherent in plastics and the like and substances absorbed or adsorbed subsequent to the container's manufacture.

All kinds of containers, including those made of glass and ceramics, can absorb odors from their contents, from cleaning compounds, from the water source, from the atmosphere, and from other contacting materials, and then release these substances into the test food or beverage (or into the gas above the sample). If feasible, lengthy heating of the cleaned container in an air oven (at temperatures suitable to the material of which the container is made – e.g., 300°F for glass or pottery) will drive off most adsorbed odorous compounds. However, heat treatment before use may actually increase the contamination from plastic containers, due to plasticizer migration and to changes in the base resin.

It is often possible to serve samples in disposable plastic cups or dishes (obviously, the container should be used only once and then discarded). The specific containers being used in the test should be carefully examined for the presence of detectable odors and flavors under the conditions of testing. As a basic precaution, the serving containers should be rinsed with distilled water (not perfect, but preferable to de-ionized water and much preferable to tap water) shortly before they are used in the test.

Consideration should also be given to the absorption of flavoring, coloring, and odorous substances from the sample between the time the food or beverage is inserted in the container and the time it is transferred or consumed. Mere leakage and mass absorption, though troublesome in other ways, should not affect the sensory properties of the remaining material, but differential adsorption can alter the relative amounts of odorants, colorants, and taste substances in the remaining sample, possibly changing the organoleptic ratings significantly.

STORAGE STABILITY

Food technologists tend to use the terms storage stability and shelf life more or less interchangeably, but most of them recognize there is a difference. Where a distinction is made, storage stability is generally used to refer to measurements of deterioration rates made in the laboratory under controlled conditions, while shelf life seems to be accepted as the term referring to the average time (or perhaps the minimum time) a product will retain good quality while being distributed under the usual conditions which prevail in trucks, warehouses, stores, homes, etc. Obviously, the latter datum will always be much less exact, but it will often be used as the basis for setting expiration dates, pull dates, etc.

The product is expected to be in good (though not necessarily optimum) condition at the printed pull date, barring exceptionally bad treatment during distribution and display, but no one can predict what consumers will do once they get the product into their homes. Milk will nearly always be drinkable on the expiration date printed on the carton when handled by the usual methods of distribution and sale, but it may sour in the consumer's refrigerator before that date if mishandled on the trip from the store or in the home. In spite of all the uncertainties in the determination of storage stability, the launch team must know the approximate shelf-life of a product before is introduces the item to the trade.

It will often be impractical to test every product variable which affects storage stability, but it is highly desirable to test every variation which experience or the literature suggests has an effect on shelf life. The product actually marketed should have been completely tested under presumed conditions of distribution and storage before it is released for sale. This means the sample subjected to the test should consist, ideally, of the food as it will be formulated and processed and when packaged in a container having the exact specifications that will be applied to the commercial product.

There are times when management makes a decision to assume that minor changes in food or package made at a late stage will not invalidate earlier shelf life studies, in order to prevent delay of a launch while additional storage studies are being run. Although such assumptions are theoretically undesirable, they may be necessary in order to meet marketing and sales scheduling. Products released in this manner must be followed very carefully throughout their distribution, and stocks should be held to an absolute minimum consistent with marketing requirements. A contingent plan should be available for withdrawing and recalling product if an unexpected defect in storage life is observed.

There are many different types of deterioration which must be taken into consideration when making storage stability studies. Among these are:

1. Breakage due to physical abuse during distribution and handling. If the product is fragile, it must be protected to the extent necessary to keep it in a form acceptable to the consumer. Deterioration of this type is related only indirectly to time, but is highly influenced by number of transfers the product undergoes and the distance it is transported. Since these two factors can seldom be controlled, at least with any great reliability, prevention of fragmentation is primarily dependent on proper design of the container.

2. Irreversible changes in texture and appearance that occur relatively quickly when product temperature reaches a critical level. The principal examples of these types of changes are chocolate products, such as chocolate coated cookies, which depend for their characteristic appearance and eating quality on the crystal structure of cocoa butter. Holding the product for even a few minutes above the critical temperature (which may differ for different kinds of chocolate) will lead to changes substantially destroying the marketability of the item, even though it may taste the same, have the same nutritive properties, and not not be unwholesome in any way.

The only preventative is to store and distribute the product at temperatures low enough to prevent these changes. Proper container design, including case design, may help to delay heat transfer enough that transient temperature changes, such as might be encountered during transfer from truck to warehouse, do not damage the product.

3. Interchange of gases with the environment.

(a) Loss of aroma or absorption of off-odors from the environment. Can be reduced or prevented by proper choice of container.

(b) Changes in moisture content. Drying due to loss of moisture is the cause of unwanted texture changes (hardening, toughening) in many food products, e.g., breads, rolls, cakes, etc. Moisture absorption is detrimental to crackers, pretzels, and various other snack products normally packaged at 3% to 6% moisture content and expected to have a crisp texture. Proper choice of container can solve these problems, but an ideal container may not be affordable.

It is worthwhile noting that plastic films which have low moisture vapor transmission rates may allow rapid transfer of the kinds of organic molecules which have strong and unpleasant odors. Moisture vapor transfer can cause a decrease or increase in water content even though the product is frozen – e.g., freezer burn, a color change due to moisture loss, is a common problem in frozen meat. Moisture vapor transfer between atmosphere and product can be prevented by packaging the foodstuff in hermetically sealed containers such as glass jars or double-seamed metal cans. Satisfactory protection can usually be obtained by sealing the food in pouches made of laminated films containing at least one layer of aluminum foil of about 0.00035 inches thickness.

A moisture resistant outer package will not, of course, prevent the exchange of water vapor between different components of a product, e.g., between the marshmallow and the base cake of a chocolate-coated cookie.

4. Gradual deterioration due to chemical, physical, or microbiological reactions taking place within the food at temperatures normally encountered by the product. Some examples of these types of deterioration which affect bakery products, as well as other types of food are:

(a) Rancidity development resulting from oxidation of fatty acids or from lipolytic reactions (usually enzymic) involving fats. Lower temperatures and (usually) lower moisture contents can reduce the rates at which these reactions occur, although very low moisture contents may accelerate the development of oxidative rancidity. Oxidative rancidity development can be slowed down by reducing the oxygen concentration in the "empty" space within the package or by including antioxidants in the food. Sometimes antioxidants are applied to the interior surfaces of the packaging material. Various types of oxygen-scavenging packets are available for including in the package.

(b) Starch retrogradation. This physical change, somewhat analogous to crystallization, causes the texture of products such as bread to deteriorate fairly rapidly – within hours in the case of certain kinds of bread, especially lean formula types. It can be delayed, or offset, by adding softeners to the formula, or by freezing the product. Refrigerator temperatures, however, accelerate the change. Low-moisture baked products, such as crackers, do not undergo starch retrogradation.

(c) Non-enzymic browning. A reaction between certain peptides and certain reducing sugars that is accompanied by off-flavor development as well as the color change indicated by the name. Mostly affects low moisture foods subjected to relatively high storage temperatures; crust color development during baking is one of the few desirable manifestations of this reaction.

(d) Microbiological reactions are serious problems in all moist foods except those commercially-sterile products which are packaged in hermetically sealed containers. Frozen foods will remain free of microbiological storage deterioration as long as they are kept solidly frozen. For non-sterile, non-frozen products, growth of molds, bacteria, and yeasts can often be controlled by manipulating the water activity, pH, salt content, and oxygen tension or by adding chemical preservatives, especially if the initial contamination is kept very low. Bakery products often become inedible by staling before microbiological spoilage occurs, although mold growth can sometime be observed on bread after three days if conditions are optimal for these organisms.

Special problems may apply to particular kinds of foods, and all deteriorative reactions must be considered before the product is deemed satisfactory for introduction. As Hegele (1988) points out, the effects of age, temperature abuse, extremely humid or dry conditions, and other physical influences are both interactive and cumulative. The technologist must learn to accept his inability to control or predict all the conditions encountered by products in the chain of distribution, and do as much as possible at the factory level to assure a stable product. Even so, it is almost inevitable that some spoilage will occur because of mishandling at the warehouse, store, or consumer level. Consumers have a knack for thinking up ways to abuse the product which laboratory people seldom imagine.

OPERATIONS RESEARCH

Operations research is a formal means of analyzing business problems. It has been around for quite awhile, but still has some lessons to teach us. The traditional instrument of the technique is a team made up of scientists from various disciplines, with an operations research specialist as the leader. Each member contributes the specialized knowledge of his field as well as any experience he has had with theories and models that might help solve the problem. Operations research has been used in planning the development and introduction of new products, although evidently not to a great extent in the food industry.

One widely used operations research tool is "critical path analysis" or "network analysis," which can be used in project planning and control to determine which activities must be done in series (i.e., one step must be completed before the next step can begin) and those that can be done in parallel (two or more steps can be worked on at the same time). These systems characteristically make use of a diagram or chart showing the temporal relationship of all the different tasks that make up the project.

Graphic representations or such types may be called PERT (Program Evaluation and Review Technique) diagrams, Gantt charts, etc. Information contained in the charts or in the data used to construct them can be compiled in tabular form for computer analysis. Using these charts, or computer models related to them, the analyst can oftentimes find the scheduling arrangement leading to shortest time through the network. This path determines the time expected to be required for completion of the project when all parts are scheduled in the proper sequence, and it is called the critical path.

An example of the use of charting methods for controlling new product development is Raytheon's "Product Planning Monitor Schedule," which is essentially a Gantt chart giving target and performance dates for all the steps involved. The chart is revised on a monthly basis using the most recent inputs of information.

Although some dramatic successes have been claimed for these procedures, they do have some defects. For example, all steps are considered as indispensable, or of equal weight, when in fact it is well recognized that some stages or components can be eliminated or condensed if management is willing to take certain risks or accept an inferior product, i.e., critical path analysis does not necessarily identify critical operations. Furthermore, all time estimates will be in error by an amount related to the predictive ability of the estimator, and these errors tend to be cumulative. Projections of technical achievements cannot be accurately made until some preliminary work has been done, in many instances, and this leads to much uncertainty.

As long as its shortcomings are recognized, critical path analysis can be a worthwhile tool in the planning of development projects. The research administrator should have at least a superficial knowledge of these techniques not only because they may be helpful in guiding research projects and integrating them into the overall company activities, but also because network analysis may suddenly be introduced by management with disruptive effects on R&D customs. A review of considerable value is the book "Critical Path Analysis" (Lang 1970).

OTHER METHODS FOR ADMINISTERING EXPERIMENTS

It is not always necessary to keep the experimental work within the product develoment department. Often, it is more efficient in terms of resources to farm out part of the work. A few of the alternatives will be discussed under this heading.

Evolutionary Operations

The technique of evolutionary operations (EVOP) is of special value for product improvement, cost reduction studies, and other investigations where radical changes in product or process are not expected. The great advantage of this approach is that operating production lines are used to conduct the experiments, so that the difficulty in translating laboratory findings to production practice, which seems to be especially great in many food manufacturing plants, can be reduced. Although EVOP is intended to be applied to existing products and processing lines, it may have some value in new product development by allowing the correction of flaws observed early in the marketing process without interrupting the product's introduction into the distribution system.

The basic principle of EVOP is the making of changes during regular production that do not alter the finished product sufficiently to cause it to be unmarketable. Changes are made in a systematic and predetermined way, and the results are carefully recorded and analyzed statistically, thus differentiating this method of investigation from the more more less intuitive changes made by line operators to improve machining characteristics or product quality during day-to-day operations.

EVOP can be applied readily to a two-variable situation, and the statistical procedure on which such experimentation is based, i.e., the mapping of response surfaces, can be applied to situations of even greater degrees of complexity. At any given time or place, however, EVOP tests should be limited to optimizing one or two variables.

In theory, the EVOP technique facilitates a constant probing in all directions to find a set of operating conditions or combination of ingredients that will result in an improved product, less scrap, higher production rates, etc. If the time and temperature of processing are the factors under consideration, a continuous study can be undertaken using the present time and temperature as the central point and making variations in all possible directions. This naturally leads to a design where the number of time-temperature variations which can be tried are five: (1) current procedure; (2) higher temperature and shorter time; (3) lower temperature and shorter time; (4) lower temperature and longer time; and (5) higher temperature and longer time.

Formula changes can be made by following a similar protocol, and such studies are particularly useful in finding the lowest possible raw materials cost for a product of given acceptability. Of course, all changes in finished products offered for sale must be within the constraints imposed by label statements and other product claims.

The EVOP technique has been described in considerable detail by Kramer (1965). He implies that EVOP studies are best conducted by the quality control group, but other authorities incline to the view that the product development section is better suited to this work.

A word of caution is appropriate at this point. It is possible to progress from an article with a high level of acceptability to an article of very poor acceptability by a taking a series of steps each of which causes a change that is imperceptible by any known or practicable means of sensory testing. Each of a series of changes can be selected so as to yield a product that an expert or consumer panel cannot differentiate at a significant level from the preceding modification, but the total effect of several of the changes can result in a final version that is not well received by the consumer. Formula alterations for the sake of cost reduction can result in loss of market through this sequence of events if the results of each successive change are compared only with the previously slightly modified product.

For this reason, it is very important to use the original, commercially successful product as the standard rather than to compare a new version with the immediately preceding modification.

Evolutionary operations can be recommended as a possible way of conducting development of line extensions, when the product changes are not expected to cause significant changes in the processing response of intermediates. For a successful EVOP program, it is absolutely essential that the willing cooperation of the Production Department be obtained. If this cannot be relied upon, it is best to forget about the evolutionary operations method of product development and product improvement.

Consultants and Contract Research Institutions

It may be economically justifiable to use contract research institutions to provide specialized assistance in a development project that seems to require technical work beyond the available capacity of a company's internal staff. These organizations differ from so-called "new product specialists" in that their emphasis is on scientific and technological studies rather than marketing or idea generation per se.

Also helpful on occasion is the individual specialist with much expertise in a technical field that is critical to a particular project. Countless hours of literature research and much fruitless experimentation can frequently be saved by resorting to the consultant's expertise. Such assistance is particularly valuable when a company is considering entry into new fields. At the start of a project, a consultant who is an expert in the particular technology under consideration can be called in to advise on possible approaches. If competent, he will often be able to point out deficiencies in some of the initial proposals drawn up by R&D. He may have specific knowledge of how similar problems were solved in other companies. Since the consultant is, presumably, acquainted with the history of litigation on products of the type being contemplated, early insight into legal considerations (and particularly the patent situation) can be obtained.

It may be thought desirable to hire one or more consultants to work alongside the R&D staff on a particularly difficult or recondite problem – temporary help, in other words. The wisdom of this arrangement is questionable; if the present staff is adequate in number and skills, it should be able to handle the details of their assigned development work. If the personnel are lacking in number and/or skills, the preferred solution is to hire more and better permanent staff members, rather than to rely indefinitely on consultants. The consultant's time is generally (but not always) considerably more expensive than that of the salaried technologist.

A consultant may be resorted to during the course of an investigation if a stalemate seems to have been reached, and may be helpful in resolving internal disputes about a product's feasibility, etc., but the consultant can hardly be expected to be a better judge than company personnel of the compatibility of a proposed new product with company facilities and staffing. If the company's executives can't make this judgement, they have inadequate knowledge of their business.

Contract research organizations can be used for projects requiring special talent or equipment, particularly in fields outside the client company's usual fields of interest. Contract research is performed by for-profit corporations, nonprofit institutes, and universities. By contracting for outside development work, increased costs for staff and investments for facilities may be reduced. Some of the benefits that have been said to result from the use of outside research groups are that it provides new viewpoints, resolves internal conflicts, relieves heavy in-house work loads, gains services of specialists, and saves time. Multi-client studies, in which several companies jointly sponsor a program on a major problem of mutual interest, are sometimes of value, but antitrust implications must be considered.

Among the reasons for hiring consultants or outside labs are (Anon. 1989):

(1) Expertise. Good consultants have specialized knowledge that a company's own employees may lack.

(2) Independence. As outsiders, consultants offer independent views.

(3) Impartiality. Because they are, presumably, not part of the company's internal policy conflicts, consultants can often serve as arbiters to settle disputes.

(4) Focus. Free from operating responsibilities, the consultant can analyze the client's problem in depth.

(5) Creativity. Consultants are less constrained by considerations of practicality.

(6) Anonymity. The company may need to pursue early stages of a project without being identifiable. This may be due to a desire to avoid tipping off competitors to the goals, or because the idea or project has some aspects that might reflect adversely on the company in the public's mind.

(7) Rapid response. Consultants can, in some cases, respond quickly to requests for data or other information not available through a client's internal sources.

(8) Reduced costs. Many highly skilled and experienced executives have been released from their positions in recent years and can often provide valuable help at much less cost than internal staff.

(9) Cachet. Some companies use prestigious consulting firms to gain popular approval for unpopular actions or radical restructurings.

(10) Multiclient surveys. If several companies commission a costly survey, the per-company cost can be drastically reduced.

Costs of a project handled by a research institute may be some negotiated firm figure, but is more often set as a maximum figure with periodic billings of overhead (usually a fixed percentage of salaries), direct salaries, rebilled charges (for travel, chemicals, etc.), and profit. Administration of the project should be assigned to a single individual in the client company who will make periodic audits of progress and expenditures. Progress reports are then transmitted by this project officer to interested parties in his organization.

In setting up an outside contract for research, it is very important to specify the goals of the project as clearly as possible. Much wasted effort can result from failure to give adequate direction to the investigators. Frequent detailed reports should be insisted on, and these should be supplemented by visits to the research facility for informal discussions with the technologists doing the work.

If difficulties occur in locating a qualified consultant or outside laboratory, information sources include directories of consultants, trade associations, your suppliers, suggestions by other staff members and consultants in related fields, and advertisements in industrial and scientific journals. The Institute of Food Technologists publishes a directory of consultants in food development that can be useful. An article by Duxbury (1993) lists about ten new product advisory organizations.

When candidates have been identified, confirm that they actually have experience in the field of interest to you and, if laboratory or other resources are needed, ensure that the consultant has access to such resources. Make certain that the consultant does not have business relationships with a competitor. Have the consultant sign a rigorous non-disclosure agreement and, if he objects, don't hire him.

When one or more possible consultants have been located, the following steps are indicated (modified from Ranger 1979):

(1) Define your problem as completely as possible.

(2) Obtain a written proposal stating what the consultant intends to do, the objectives, the time required to complete the task or reach a review stage, the costs, and the way results will be presented to you. A definite termination point should always be defined in advance, particularly when a fixed cost is agreed upon. Recognize that assignments and objectives tend to drift or change as work progresses and be ready to adapt to the actual needs as they arise.

(3) Assign one or more technologists to work closely with the consultant. A consultant should seldom be left to work independently unless the assignment is not connected with other company activities.

(4) Set up check points or a regular report system to keep the task on track and properly coordinated.

(5) The project should end with a written report containing definite conclusions and recommendations. A meeting of all or most of the interested parties within the company should have an opportunity to discuss the report, or at least discuss the parts related to their work. Recommendations for further actions may be taken at this time.

Labuza and Lineback (1987) listed about 20 research centers in which the food science departments of colleges or universities are involved. You may be asked to give a "grant" to one of these university departments or captive institutes with the understanding that one or more graduate students will work on your project under the direction of a well-known or even a capable food scientist. Be aware that the "well-known" scientist may have many other responsibilities much more pressing than your project, so that his contribution to the solution of your problem may be quite limited. Ordinarily, you will allowed to have very limited influence on the actual experimentation and none into scheduling. If you are tempted to use services of this sort, ask how many other projects the scientist is supervising and what his results have been for **industrial** (not government) clients in the past; determine the qualifications of the graduate student(s) that will be working on your project. If you want quick service, go elsewhere.

Be aware that most consultants, upon completion of their assignment, will suggest that further work is needed to arrive at a final answer; work requiring their participation, of course. This is just good business on their part, and is not unethical, provided they have not deliberately suppressed conclusive results that would have fully answered your questions. Decide whether or not they have given you value for money in the past before you decide to continue the relationship.

Enlisting the Help of Suppliers

If your company is fairly large and you are in a position to influence the purchase of supplies, it is usually possible to get a great deal of technical help at no cost from ingredient and equipment manufacturers who hope to sell components of the new product or the machines needed for its production. Relationships that are mutually profitable can result, and they need not be unethical or in any way detrimental to the quality or cost of the final product. It is important to keep in mind at all times that the relationship must be an arms' length one; your own company's welfare must be taken into account before any favors are granted to the supplier.

Sources of ingredients and materials can be found listed in the annual directories of such magazines as Food Processing, Prepared Foods, and the like. A problem with these directories is that repackagers, distributors, and other secondary sources are listed alongside manufacturers. It is rather rare to be able to get reliable technical assistance from any company that is not a primary source. Whether or not the technical services department of a primary source is willing to devote any time to helping an unaffiliated individual with an idea or a small company with limited prospects depends upon the individual contacted. It may be more profitable for such entrepreneurs to talk to a lower level employee in the lab of the primary source, instead of a sales representative, if a contact can be made through mutual friends or otherwise. Do not be modest about your new product's prospects when requesting help from salesmen or higher level executives of the supplier; be optimistic when estimating how many million units you expect to sell annually when your product has been in full national distribution for a couple of years.

For ingredients that are hard to find, a software program called "Ingredient Findex" is said to enable rapid computer searches for sources. The developer says the program "enables you to conduct a fast confidential search . . . [in a] database that contains more than 21,100 ingredients from 1,100 suppliers." A search can be conducted using up to ten descriptive words, such as the ingredient name, its applications, how it functions in a finished product, the product's trade name, or the supplier. Disks are updated every 90 days. Ingredient Findex is distributed by Hopkins Technology. The author of this book has not had an opportunity to review this software.

New Venture Groups

Some larger companies use the venture group approach to guide the development, production, and introduction of new products and new lines of products. The characterizing feature of the venture group for new products is that the personnel managing it are totally committed to the one goal of successfully introducing the particular item assigned to them. Venture groups could be considered as a more autonomous version of the "teams" mentioned several times earlier in the this chapter, but members of a team owe allegiance to their particular departments and accept primary direction from their department heads while members of a venture group are more or less isolated in their responsibility to a goal of successfully introducing into the market a single product line and they are full-time workers to that end. The nomenclature tends to get a bit confused at times.

A Pillsbury product development project was described as a team effort (Anon. 1993), but the "team" could easily fit into our definition of a "new venture group." In this project, a 20-member team was formed from Pillsbury personnel experienced in product, process development, packaging, operations, sales, distribution, finance, consumer service, quality assurance, marketing, and information systems, The group was headquartered in the pilot plant of the company's technology center. It had a considerable degree of autonomy and, correspondingly, a high level of accountability.

The Pillsbury venture group operated on a "shared leadership" principle, meaning the team was led by the person responsible for the function that was most critical at the time. Although the project was apparently quite successful under this method of direction, many possible conflicts can be foreseen as people are bound to differ in their ideas of what is most critical. Probably, as a practical matter, there was considerable oversight of the day-to-day operations by an management-appointed head of the group.

Advantages of the Pillsbury approach was cited as (Anon. 1993), "Using the cross-functional approach, the team was able to reduce product development time because the work functions could be done in a parallel manner rather than sequential manner. For example, while the recipes for the products were being fine-tuned, the shapes of the individual products were being finalized and the packaging was being developed."

Bates (1967) claimed the venture group approach was better, faster, and cheaper than other methods of new product development. It has its dangers, however, in that the enthusiasm and emotional commitment generated in the group's members by the entrepreneurial concept and their hope of reward may lead them to ignore or even to conceal substantial negative factors that may come to light during the course of the program. In addition, the very nature of the method leads to a relatively large outlay of high quality manpower and funds on behalf of a product at a relatively early stage in its development – when the product's potential, if any, has not been fully evaluated.

Venture teams may consist of one or more persons. Three seems to be a common number. Bates summarized their functions as follows. The manager maintains liaison with top management, develops and matures the concept, makes marketing strategy decisions, presents the product to potential clients in and out of the firm, and provides general administrative guidance. The marketing/economic/financial representative assists with concept development, draws up product profitability statements, conducts market research, plans marketing action, performs sales tests, does product administration, and takes charge of venture cost control. The representative of technical/research/production functions assists with concept development, evaluates technical feasibility, supervises technical research programs, develops the product and its package, makes co-packer arrangements, and oversees production through the test marketing stage.

TIMELY TERMINATION IS IMPORTANT

In order to save time and money, it is very important to terminate inappropriate concepts at the earliest stage at which seemingly insurmountable obstacles can be recognized. Some idea of the costs involved if this is not done can be gained from a review of General Foods' experience. These data are several years old, but are useful to indicate the magnitude of costs which can be incurred by large companies. General Foods, during a 10-year period, studied more than 600 new product ideas (averaging 60 per year) at a cost of about $76 million. Of this total, about $30 million went into lab work, and the remaining $46 million went into marketing expenses resulting from the testing of product concepts.

About 118 of these product ideas were deemed worthy of further development. Additional testing eliminated another 31, leaving about 87 still viable. The 87 that went into test market dropped to 40 possibles, and 10 more died after national introduction. The final 30 products represented a total investment of $243 million, an average of $8 million per product (Anon. 1973).

Obviously, new product development is costly. Each concept subjected to technical and marketing studies absorbs a very significant percentage of the company's resources. Funding must come primarily from the company's profits (if any), although some tax benefits should accrue to partly offset this drain. If the company is not profitable, the added expenditure must come from capital – perhaps, ultimately, from the company's creditors or shareholders. There is a very good reason, then, for reducing the number of concepts that must go through any sort of laboratory or marketing studies. This means eliminating at the earliest possible stage, after a minimum amount of actual work, those concepts which do not have a nearly perfect fit with company policy, or about which the technical and marketing group express serious reservations based on a consideration of the fundamentals.

BIBLIOGRAPHY

ANON. 1973. The rebuilding job at General Foods. Business Week (Aug. 25), 48-55.

ANON. 1976. From concept to product – in 21 steps. Plastics Eng. *32*, No. 7, 1-4.

ANON. 1989. Chemical industry consultants. Today's Chemist *2*, No. 2, 13, 15-16.

ANON. 1993. Brand leverage. Baking & Snack *15*, No. 2, 102.

BALDERSTON, J. 1976. The research budget. Master or servant? Res. Dev. *27*, No. 8, 52, 54.

BATES, B. 1967. Venture group approach to new products. Food Prod. Dev. *1*, No. 5, 34-42.

BEST, D. 1986. Assesing the risks of research and development. Prepared Foods *155*, No. 10, 150-152, 156.

DREIER, W. 1988. Avoiding the common problems of process scale-up. Prepared Foods *157*, No. 6, 131-132.

DUNNIGAN, J. A. 1988. Bringing your product to life. Part I. Patenting your product. Entrepreneur *17*, No. 3, 58-61.

DUXBURY, D. D. 1993. Let's send out for R&D. Food Processing *54*, No. 5, 30-32, 35, 37.

DZIEZAK, J. D. 1987. Technical centers expand product development capabilities. Food Technology *41*, No. 3, 110-111.

EPNER, S. A., and BADER, J. 1984. How to make the most of consultants. Today's Office *19*, No. 2, 29-31.

HEGELE, F. A. 1988. Shelf life. Cereal Foods World *33*, 237-238.

HILTON, P. 1971. Exploit new product technology. Food Prod. Dev. *5*, No. 5, 38, 40.

HOPKINS, D. 1980. Survey finds 67% of new products succeed. Marketing News *13*, No. 16, 1.

HORN, W. 1983. Economics of purchasing new equipment. Proc. Am. Soc. Bakery Engineers *1983*, 124-131.

JACKEL, S. 1980. Systematic in-bakery research using EVOP methods. Proc. Am. Soc. Bakery Engineers *1980*, 74-79.

JOGLEKAR, A. M., AND MAY, A. F. 1987. Product excellence through design of experiments. Cereal Foods World *32*, 857-868.

KONING, J. W., JR. 1974. The Scientist Looks at Research Management. Amacom, NYC

KRAMER, A. 1965. The effective use of operations research and EVOP in quality control. Food Technology *19*, 37-39.

KUNTZ, L. A. 1991. Accelerated shelf-life testing: Is faster better? Food Product Design *1*, No. 12, 42-43.

LABUZA, T. P., and LINEBACK, D. R. 1987. The university-industry relationship in food science and technology. Food Technology *41*, No. 12, 74, 76, 78, 80-91.

LANG, D. W. 1970. Critical Path Analysis. Dover Publications, NYC

LAWLESS, H. T., and CLAASEN, M. R. 1993. Application of the central dogma in sensory evaluation. Food Technology *47*, No. 6, 139-146.

MARTENS, A. E. 1984. Make sure your R&D is as good as it can be. Res. Dev. *35*, No. 11, 86-88.

MILLER, D. B. 1986. Managing Professionals in Research and Development. Jossey-Bass, Inc., San Francisco, CA.

MOLDAWER, R. A. 1976. The marketing research steps that aid sound product development. Package Dev. Syst. *6*, No. 5, 40-42.

MONS, J. 1987. Pilot plants bridge new products and production. Prepared Foods *156*, No 8. 175, 178, 180.

O'CONNOR, P. 1993. Managing product teams. R&D Mag. *35*, No. 8, 67.

O FAIGAN, C., SHEEHAN, H., and O'KENNEDY, R. 1991. Accelerated degradation testing. Am. Biotechnology Lab. *9*, No. 9, 31-32.

PEARCE. S. J. 1987. Quality assurance involvement in new product design. Food Technology *41*, No. 4, 104, 109-110, 113-114, 116.

RANGER, H. O. 1979. Outside consultants: How to use them successfully. Package Development & Systems *1979*, No. 2, 6

SACCHAROW, S. 1986. How to measure shelf life in food packaging. Prepared Foods *155*, No. 7, 93, 96.

SCHENGRUND, D. M. 1983. Managing new product technological services. Manufacturing Confectioner *63*, No. 5, 58-59.

SCHRELLO, D. M. 1975. Twenty questions for product decisions. Industrial Res. *17*, No. 9, 83-84.

SEILER, R. E. 1965. Improving the Effectiveness of Research and Development. McGraw-Hill Book Co., NYC

SOUDER, W. E. 1970. Pitfalls in R&D budgeting. Management Review *1970*, No. 6, 54-56.

SULLIVAN, D. 1976. Good product concepts embody business goals, market needs. Food Prod. Dev. *10*, No. 4, 28, 30.

STANLEY, A. O. and WHITE, K. K. 1965. Organizing the R&D Function. American Management Association, New York.

THOMPSON, D. R. 1982A. Production optimization. J. Food Processing Preservation *6*, 1-13.

THOMPSON, D. R. 1982B. Response surface experimentation. J. Food Processing Preservation *6*, 155-188.

WALTERS, J. E. 1965. Research Management: Principles and Practice. Spartan Books, New York.

WHITE, V. P. 1988. Handbook of Research Laboratory Management. ISI Press, Philadelphia, PA

chapter seven

INTERFACING

INTRODUCTION

Interfacing, as the term is being used by management theorists today, means the contacts occurring between associated groups (or even individuals) who are, or should be, working toward a common goal established by a company or some other organization. This definition has to be regarded as tentative and approximate considering the nature and status of management theory. Generally, this term is not taken to mean contacts with persons in the direct chain of command above the individual or with persons directly supervised by an individual. Sometimes (as in the following discussion) interfacing may be regarded as including contacts with persons or groups outside of the company.

One of the many factors having major effects on the efficiency of product development systems is the functioning of the interfaces between the different groups who must perform the various actions needed to complete product formulation and process design as well as all the related measures involved in getting a product from concept status to the mass production stage.

In addition to the technical work being performed by a nucleus group in product development, there is required a flow of information between those departments of the company which are required to participate in the project, an input of data and directives from government agencies and other outside participants (wanted and unwanted), the availability of funds required for implementing the necessary actions, and the frequent involvement of top management. In a product development project, everyone who should be involved must be involved, and all others should be kept out of the way.

Much of the interfacing in a typical project will take the form of the transfer of information between different departments. Although this information is often intended to secure the performance of some action by the recipient, this is by no means always the case. The mere alerting of another department that some action which may affect it is about to take place is also a common feature of interfacing. In the case of government agencies that may have some relevance to the project, obtaining inaction may be the goal of the communication.

The smoothness and speed with which these interactions occur is partly a measure of the skills of the project leader and partly a result of the support he or she receives from top management. All the other department heads and outside contacts have goals and constraints of their own, and it is not always easy or even possible to gain their willing cooperation by appealing to their company loyalty or by emphasizing the need to improve economic prospects of their employer. Some of them may have disagreed with the original idea for the product and now have an emotional commitment to causing its failure, others may believe they already have too much to do and will try to avoid spending time on another person's project, still others may be rigid and argumentative by nature and tend to oppose all new assignments as a matter of personal policy.

The project leader must, then, be something of a diplomat. He must be able to persuade those persons whose cooperation he needs that providing the necessary assistance is in their own best interests as well as for the benefit of the company. Appeals to top management to force cooperation should be necessary in very few, if any, situations, because going over the head of another executive nearly always arouses an intense and long-lasting antagonism in the affected party which may ultimately make a successful conclusion to the project impossible.

In theory, at least, technical people tend to rely on facts, i.e., evidence, as the basic motivating force causing an individual to choose a course of action, while many of the people the technologist contacts in the course of his work will regard facts as merely minor obstacles to be overcome in establishing their own agenda. This situation mandates the preparation of project reports and requests for assistance in the plainest language, devoid of technical jargon. Graphs and charts should be used instead of tables, if that is possible. Models or mock-ups of finished packages will often be more persuasive than drawings or other artwork.

It is not a universal rule, but can be observed as a frequent occurrence, that cooperation between departments is somewhat easier to obtain at the start-up of a project (partly because the heads are motivated by curiosity and partly because relatively minor drains on personnel and facilities are required at this stage); assistance tends to become more difficult to obtain as the project drags on (particularly if no major advances are visible to the non-technical participants); and, finally, in the "polishing up" phase of the project when the product appears in a recognizable commercial form, other departments become more enthusiastic and cooperative as well as more determined to get this irritant out of their way as quickly as possible. If the product shows unmistakable signs of becoming a commercial success, it becomes much easier to obtain help from almost anyone, most of the helpers hoping to share in the credit and trying to establish a basis for claiming they provided the most important (and probably the first) support for the project — few proverbs are truer than the one that says, "Success has a hundred fathers, but failure is an orphan."

Each department head should designate one of his subordinates as the chief contact within his group for the leader of each product development project. It is possible that one person in each department will handle all of these projects, but this is not necessarily true or even advisable. Especially in the cases of Marketing and Production Departments, it is very likely that different liaison persons will be assigned for each project, depending upon the special expertise needed for manufacturing and merchandising that particular product. It may even be necessary to change the contact person as the project reaches different stages. It is certainly necessary, however, that a contact be available in each of the Departments to ensure needed assistance to the project leader without delay. The project leader, and presumably all other members of the team, should have in their files a complete list of the current contacts who may be called upon for aid and advice on the project.

The following paragraphs cannot provide solutions for all of the human problems a product development leader may encounter in coordinating the activities of co-workers in other departments, but they may suggest points of departure in planning and executing the necessary interfacing activities.

MARKETING

Marketing people generally have a love-hate relationship with new products. They like the idea of having another opportunity to exercise their "creativity," but they are often afraid they will become associated with, or blamed for, the failure if the new product does not meet sales goals after it reaches the stores. They tend to demand unrealistic or even impossible product characteristics as a way of outdistancing competition and making their own work of introducing the product easier, and they are prone to accuse technical people of footdragging or obstructionism if the technologists point out the difficulties in achieving marketing's wish list. They have a pronounced tendency to make quick intuitive judgements of product quality and consumer expectations and do not have much patience with objective data obtained from physical and chemical tests, especially if the evidence seems to contradict one of their favorite proposals. As reported by Hegenbart (1990), one marketing manager said, "The marketing/new product introduction process demands quick introductions. This puts R&D in the position of having to turn things around faster than they would like." Also, "R&D tends to be rooted in place; everything has to be proven for them. They don't seem to have any imagination."

Somewhat different thoughts were expressed by Michael Brandt, a Campbell Soup executive who had worked in both R&D and marketing (as quoted by Mullich 1993): ". . . it is critical for marketing and R&D to work together . . . Otherwise deadlines may not be kept, project objectives may not be fulfilled . . . R&D people tend to have a lot of experience, where marketing people are quite often a relatively young group of people . . . R&D usually has some people who have been in the business a long time. They've seen a lot; they've seen the losers and winners. R&D can provide a lot of input into strategy even if they are not schooled into it from an educational point of view."

Marketing managers generally are eager for interactions with advertising agencies where they find free-wheeling creativity and bon hommie compatible with their own personalities, but they are much more reluctant to interact with technicians, whom they tend to regard as stodgy, slow-moving, boring, and unimaginative – generally square, bubble-bursting types. Or, as expressed by Frand (1984), "Marketing and sales people . . . feel that laboratory people are somewhat naive, unwordly, and really don't know what's going on. Furthermore, they have the feeling that if they were to explain to the laboratory their problems and their needs, they probably wouldn't be listened to . . . it is vital that the marketing people are allowed to have closer contact with the laboratory and be involved in the inputs at the early stages of product development."

An interesting summary of the sequence of new product development from a marketing manager's standpoint was given by Struse (1977), at that time General Manager of Marketing Research, Oscar Meyer & Co. His description of the the problems that exist is given below (considerably abbreviated, and, in part, modified):

(1) Because the manager doesn't have a reliable fix on the sales potential of his idea, he has to resort to using his credibility and expertise to his superiors and other managers. At this point, there are at least two possibilities:

(a) The manager has become emotionally committed to his idea. If his idea is not really as promising as he believed, he now has to admit he was wrong and "unsell" his idea. The net result is a loss of credibility and perceived expertise. Such a course of action might not be politically sound, especially if one or more of the manager's superiors has become a supporter of the project on the basis of the manager's recommendations.

(b) The corporation becomes committed. The nature of the commitment of the parties differs. The manager is wed to the project and to making it suceed; producing an optimum product is his goal. On the other hand, his superior may be more interested in getting a quick conclusion to the project.

(2) At this point, the project may be approved. The marketing manager is generally faced with trying to get specific tasks accomplished by other members of the organization, persons who were not involved in the creation of the project and bring their own different perspectives on what the product should be. The manager has to get them not only to understand his conceptualization of the product but to accept his commitment. If he is unsucessful in this, he can either avoid open conflict by negotiating compromises or he can allow the organization to resolve the conflicts (usually done by compromises on the original concept). In either case, the end result will usually be a product that includes inconsistent and often incompatible characteristics – almost a certain prescription for product failure. In any case, all parties involved become more committed at each stage.

(3) When all the pieces are put together, the manager has to sell the company on determining whether anyone will buy the product. Perhaps it is for-

tunate a good many products die at this stage because it doesn't appear there are too many people left who can objectively interpret the test market results.

(4) All too often the test market becomes positioned as an exercise to get some experience in handling the product or selling it into the trade or to see how high is up. Consequently, encouraging early results are often run with, while discouraging results are discounted, and test markets are extended or repeated.

Struse indicates his comments represent an exaggeration or caricature of reality, but they sound a lot like a number of experiences I have had, with no exaggeration at all.

According to Goldenfield (1977), marketing's decisions relating to new product development, especially in their role as intermediaries between product development personnel and advertising agencies, include (among others) the following:

(1) The marketing positioning of the product and the advertising strategy.

(2) Finalization of the product name, including both the trade or fanciful name and the product identity or common name.

(3) Completion of advertising copy for television, magazine, coupons, radio, and newspapers.

(4) Determination of label art and label copy, as well as final package design.

(5) Agreement on what advertising claims will be used.

In contacts with marketing people, the technologist must always present a cooperative posture but be wary of getting caught up in all of their enthusiasms. Never adopt a negative attitude toward their product ideas (unless there is a compelling reason for doing so) but do indicate that Marketing's ideas must go through the same evaluation steps as every other suggestion, "as a matter of company policy." This prevents the accumulation of a large number of unofficial projects that clutter the system and divert energies from prioritized products.

It is definitely not wise to prepare special prototype samples for Marketing's evaluation unless the samples have been made in accordance with standard procedures and all details of the preparation and presentation have been fully recorded. If these precautions are not followed, you may at some later time be requested to duplicate "exactly" a sample that the evaluators remember only vaguely; if at all possible, keep several duplicates of management review samples in storage indefinitely, and, in every case, record all significant circumstances surrounding the steps of producing and examining the sample. Never submit for examination a product sample that does not bear an identification number and, preferably, do not submit one without an accompanying memo containing the identification number(s). Delivering back-door, out-of-the-system samples is dangerous practice, as well as being time-consuming and unproductive.

During that late stage of development in which refining or fine-tuning of product characteristics takes place, Marketing will undoubtedly be conducting a series of consumer acceptability tests. If at all possible, the team leader should assign an observer from the technical development group to attend each of these test sessions. Very serious errors can be introduced into the test results by poor practices in presenting the samples (allowing temperatures to vary, for example), in collecting data (discarding or losing some of the ballots), in failing to properly identify samples and/or panelists, or in giving incorrect instructions on test methodology to the subjects. These discrepancies may not be reported or even noticed by persons conducting the test, especially if an outside agency is in charge. It is even possible that the data may be deliberately "cooked" so as to make it support a conclusion preferred by Marketing, human nature being what it is. This fudging may take the form of using unorthodox statistical analysis methods, particularly that of running additional tests with different subjects when the original results are unfavorable.

Laster (1976) described a technique for improved R&D and business planning called the "portfolio concept." It assumes that the business needs for a given product or product category can be summarized along one side of a single sheet of paper, while the technical programs designed to meet those business needs can be lined up on the other side. Thus for a product or category, the business/technical interface is reduced to a line running down the middle of the page. With information displayed in this way, it becomes easier for people whose primary interests are on one side of the line to know what is going on on the other side and to mesh and balance their efforts with work being done on the other side.

FINANCE

The one department regarded by Marketing's "creative" geniuses as being more of a dream-killer than the data-oriented R&D people is Finance. It is true that many accounting types tend to have a very negative, even obstructive, attitude toward expenditures not resulting in some immediately detectable benefit. They often demand "proof" that a new product will bring in more money than it will cost to develop. Of course, such proof is never possible, since too many variables and too many unknowns are always involved in developing and introducing a new product. As a result, some product development teams will, in desperation, feed Finance data that have been obtained from highly suspect sources and fail to transmit information that might have some bearing on the problem (i.e., certain alternative choices that might be available). The CEO must furnish proper guidance to Finance or himself assume responsiblity for decisions that are not justifiable on a 2+2=4 type of rationalization if he expects to have an efficient new product program.

The correct role of Finance is to furnish the most reliable estimates of new product costs based on information given them by other department heads, not to give a yes or no decision based on some hypothetical yield based on assumptions that are truly unknown, such as interest costs in future years or trends in ingredient prices. Can the company afford such investments? That is not a question Finance needs to answer, it is the ultimate responsibility of the stockholders as represented by the Board of Directors, or, in practical terms, the CEO.

According to Westwood (1988), "The secret in handling the R&D/ finance interface is to let the finance people in on the excitement. Invite them to the laboratories and take the time to explain, in plain English, what's going on, and why. Let their imaginations loose on the latest breakthrough, and ask their advice on how to bring it to market. If R&D has done a good job of communicating, they will say: "Why don't we do do so and so? . . . My experience is that finance managers can be extremely creative regarding ways of funding projects they consider exciting, and can be as daring as reasearchers when they have a good feeling – not necessarily an understanding – about what is going on, and have been encouraged to consider themselves part of the team."

PRODUCTION

Even though the head of the Production Department may express enthusiasm about, and willingness to cooperate with, product development activities, he will almost always be reluctant to begin a new project. He can foresee the interruptions in his manufacturing lines as sample runs are scheduled, he knows he will be saddled with the responsibility of producing uniform new products in a cost effective manner without sufficient time to learn all of the possible problems that can occur during extended runs, he will expect disruptions of the plant as specialized equipment is installed to support unique processing steps that are found to be required, and he will be faced with the disturbing presence in the factory of individuals and teams from other departments.

Overall production efficiency will almost certainly decline during test runs and process development, and top management may be disinclined to to accept Production's excuse that new product development was the cause. If test runs fail to produce acceptable products, the Production Manager knows that – rightly or wrongly – his department will receive some or all of the blame. On the other hand, if an acceptable product finally comes to fruition, the Production Manager rarely gets his fair share of the credit.

It is generally true that the Production Manager in most companies has more day-to-day problems requiring his immediate attention than any of the other managers in the firm (due in part to the large number of people under his control), and this constant pressure induces a strong tendency to avoid additional potential problems. The development manager should understand this situation and make a special effort to be flexible in scheduling test runs and the like.

The production manager's problems can be greatly alleviated if adequate budget allocations are made to his department to compensate for the lost personnel time and reduced line capacity resulting from pilot plant and production-scale work on the new product. Of course, exact costs are very difficult to predict, but the manager's attitude will be much less favorable if he has to struggle with the Accounting (or Financial) Department for adequate recompense after each trial run. On the other hand, the project leader must audit the production manager's charges to ensure that costs for the project are not being inflated as a means of improving Production's financial standing.

If the company is fortunate enough to possess a pilot plant that can be used to scale up laboratory batches to semi-plant sizes, the problem of bringing test formulas and procedures into full production is considerably ameliorated. In an ideal world, the pilot plant would be efficiently managed by a consortium of R&D, Engineering, and Production managers, but failing this utopial situation, it is probably best managed by Engineering (provided the Engineering group is not under the direct control of Manufacturing), or, next best, by R&D.

Direct scale-up costs are nearly always considerably reduced by pilot plant tests, and indirect costs related to interruptions of Production's normal schedules are always greatly reduced. Scrap production should also much less when pilot plan runs are used to rough out the process conditions and formula specifications. Positive results in pilot plant batches can help to reduce the gridlock that develops when a Production Manager uses manufacturing problems that are discovered in early runs as bases for claims such as, "We can't possibly run this stuff efficiently."

The Production Manager should be kept constantly informed of the status and plans of the product development team. During the early stages of the project, it is probably not necessary that he attend every meeting — in fact he will probably be unable to attend most of them — but his presence in a meeting will definitely help to improve the flow of information and avoid unpleasant surprises. It is not always possible to predict what he will consider a major stumbling block, and these potential obstacles must be identified with minimum delay.

In the event of prolonged difficulties in developing new production lines (or modifying existing plants) to manufacture the new product, the management of Production and Engineering sections will be certain to demand that the formula be changed to make the mixture easier to process. Such changes may, in fact, become inevitable, but it is essential for the project manager to point out that changes affecting finished product characteristics (including advertising claims) must be cleared with top management as well as with the technologist. In most cases, the easiest way out of a production problem is a process change, not a formula modification, and a process change rarely leads to the need for a labeling change.

QUALITY CONTROL

If the Quality Assurance group or section is directed by the same scientist that guides research and development, close coordination of the two groups is virtually assured, but, in any event, relatively few tensions should develop as product development proceeds in its path from idea to product.

The new food and its package may require materials not currently being procured by the company. The QC department must be alerted to this problem as soon as possible so that specifications, and the test procedures required to enforce them, can be drawn up as soon as possible. If a similar ingredient or packaging material is currently being procured, existing specifications must be reviewed to ensure that the limits for acceptable deliveries are compatible with the needs of the new product – the company may be buying "flour" at the present time, but does this "flour" have the properties that qualify it as the "flour" needed as an ingredient in the proposed new item?

Specifications that have not been used for some time, as for an ingredient that had been procured for a product no longer being made, should not be pulled out of the files and put into effect for the new product without a thorough evaluation of their current applicability – with respect to availability, legal status, cost, etc.

The concept of having product design and feasibility validated by a continuous review of specifications during development, has been called Design Assurance. Further details can be obtained from "Food Processing Industry Quality Systems Guidelines," published by the American Society for Quality Control. As summarized by Pearce (1987), the guiding principles of a design assurance policy should be (slightly modified):

(1) Design work is to be done in conformity with requirements defining the marketing concept, consistent with regulatory requirements.

(2) Design work is to be done by proper procedure and standards.

(3) The results of design work are to be documented and changes controlled.

(4) Designs are to be proven by test at each stage of refinement (scale-up) and after full-scale production trials.

(5) Design work is to be formally reviewed for effectiveness by third-party experts at milestone stages and problems corrected before the design is released to the next stage.

(6) Planning and coordination of all activities needed to support new product manufacture are essential.

The finished product must also be described by a full set of specifications as well as protocols for the tests used to determine the critical factors. The ranges or limits of the qualities to be tested for must be established by consultations of the technologists in QA and product development joined by representatives of Marketing and Production. Marketing will have to decide how much the product can differ in its consumer-detectable characteristics without affecting repeat sales, and Production will have to decide upon the feasibility of manufacturing lines that can consistently reach the proposed targets during normal production runs.

Product development must inform QA promptly of all changes made in the food and package that influence specifications and testing. Otherwise, unacceptable delays may occur in finishing the project, because it is very unlikely that management will approve the manufacture and sale of a product for which there is not a reliable means of assuring the presentation of uniform examples to the consumer.

PURCHASING

Purchasing generally is assigned the responsibilities of locating suitable suppliers for ingredients, containers, (possibly) equipment, and certain other items required for manufacturing the new product. The project leader must work closely with the purchasing representative to ensure that necessary actions are being taken.

The technologist may be able to modify the initial specifications for materials so as to give the Purchasing Department a greater number of potential suppliers, thereby increasing competition (with the possibility of decreasing materials costs and freight costs) and simplifying delivery scheduling.

Purchasing personnel will have to receive sets of tentative specifications for materials as soon as possible after they are written, and they must be informed immediately of any changes in these specifications. Purchasing must inform the project leader of any unresolved problems in locating suitable suppliers for an ingredient or packaging material. The technologists may be able to assist in locating suppliers, especially for recently offered materials that may not yet be listed in the catalogs and other references being used by the Purchasing Department.

Occasionally, an aggressive purchasing manager will attempt to convince a team leader to substitute a cheaper ingredient for the one the technologists have been working with. Never accept his word that the substitute is just as good; he can't possibly know this is true. Usually, he is taking the word of a favorite sales representative from a supplier. If you let the purchasing manager (or the manager of finance) dictate your formulas and processing methods, you run a high risk of specifying a food which the plant can't produce, your customers won't buy, and the government won't let you sell.

DISTRIBUTION

In many companies there will be a separate department or other subdivision bearing the title "Physical Distribution," or something similar. and it will be charged with the responsibility of controlling products from the time they enter the company warehouse until they enter the customer's warehouse (or customer's transportation vehicle, if the customer is picking up the goods at the plant's dock). The physical distribution group also maintains the records of stocks, generally conducts (or helps to conduct) inventories, and has other functions.

There will be many occasions during a product development project at which the manager of the project must interact with the Physical Distribution Department. In very large companies, the head of physical distribution may wish to delegate a full-time representative to interface with the product development group, with the assignment of maintaining liaison on all points of mutual interest and securing action of persons within the physical distribution department when necessary to advance the project. In most companies, however, the physical distribution department will not have enough available personnel or funds to assign a full-time representative to the project, and must instead allocate only part of the time of one or more individuals for this purpose. In any case, frequent memos describing the needs of the product development group should be forwarded to the head of physical distribution, with interdepartmental conferences held as necessary to explain special requirements in more detail.

The physical distribution representative must be kept informed of any special storage requirements of projected new products, their storage life at various temperatures, their susceptibility to mechanical and other damage during transportation, the approximate quantity (number, weight, volume) of the new product that will be entering the warehouse at various times, and any special routing requirements for sales test materials.

The physical distribution department will be called upon to provide transportation vehicles, outside warehouses, and other necessary facilities for the new product at various phases in its commercialization. For example, provisions must be made for expeditious handling of sales test material and samples that are sent out to measure the response to handling and storage of the product in its final packaged form. There are situations where small amounts of sample are best transferred informally (e.g., hand-carried) by members of the product development team without the participation of physical distribution simply because this is speedier and can be better controlled.

In cases where the product is sensitive to environmental influences, and this includes virtually all foodstuffs, the finished packaged material must be moved within the company under complete control insofar as temperature, humidity, mechanical treatment, and other environmental factors are concerned. Furthermore, each case or other shipping container must be completely identified and its location recorded at all times.

The final document describing the product must include specifications for storage, handling, and other distribution conditions; shelf-life for the product under the conditions specified must be clearly set out, and its relationship to the expiry dates to be affixed to each consumer unit should be described. Means of disposing of returned or unsold product having expired dates must be prescribed.

Physical distribution will also be responsible for obtaining delivery of, and (usually) for storing ingredients, packaging materials, and other components of a new product. Although the purchasing department is responsible for the actual buying of these items, and usually assists in (or takes charge of) locating the suppliers, the functions of coordinating delivery, having suitable means for accepting delivery of bulk materials such as corn syrup or flour, and maintaining adequate stocks of fresh materials and rotating these stocks are usually assigned to physical distribution. The leader of new product development must confirm that the physical distribution representative is fully informed at all times of any changes in the product requirements affecting these operations, even though estimates of quantities required and scheduling of their availability may originate with Production and/or Marketing.

LAWYERS AND REGULATORY AGENCIES

Assuming ever-increasing importance in the guidance of new product development is the constantly expanding activity of national and state governments; in some places, even the municipal governments get into the act. In the aggregate, these entities seek to regulate essentially all of the characteristics of foodstuffs offered for sale, as well as their packaging, labeling, advertising, and means of production and distribution. In some instances, Federal and local regulations conflict and overlap, when congressmen have not had the foresight to include pre-emption in their bill, and already onerous burdens are exacerbated.

It is absolutely essential that the persons involved in developing new products have ready access to reliable sources of information on existing regulations as well as to a source of advice on specific questions that may arise during the course of the research. It is useless and even dangerous to rely on personal opinions as to what seems to be morally right or technically effective, since laws and regulations in the food field are often based on doctrinal or political considerations that have little or nothing to do with common sense or logic, much less with ethics.

The most obvious solution to the regulatory problem is to have on the firm's staff a lawyer who specializes in food regulatory work. If such an employee exists, he or she must be kept intimately involved with all new product programs at all stages of their development. A staff attorney does not need to attend every product development meeting and he does not have to be a permanent member of the committee or team, but he should keep himself aware of all the developments in active projects by studying reports and asking for clarification when a point seems obscure. He should immediately inform the project leader if a conflict with laws or regulations is occurring or appears to be in prospect. He should also inform the leader of any pending change in the law, and provide an opinion on that change's effect on development plans.

Since only the largest companies will be able to afford a staff attorney specializing in food law, it becomes necessary for their product development teams to work with outside lawyers. Although numerous law firms have on their staffs one or more attorneys specializing in food regulatory work, it is often difficult to locate legal advisers who are sufficiently technically oriented to be able to understand the scientific problems connected with new product development. Nonetheless, assistance of this caliber is highly desirable for helping to facilitate the resolution of those stalemates which result when essential technical limitations seem to be incompatible with legal requirements.

A free flow of information between the project leader and the outside attorney should be occurring during the course of the project, even though this way of proceeding will inevitably result in substantial billings by the legal adviser. It will seldom be necessary for the outside attorney to attend product development meetings, and (for security reasons) it may not even be desirable to have him visit the plant. Discuss legal matters with the attorney, don't bother him with ethics: a lawyer has the same attitude toward ethics as a shark does toward vegetarianism (little knowledge and no interest).

Geller (1977), writing on the topic of interfacing between a company's legal advisers and product development personnel, stated that the following must be taken into account when considering the establishment of a new product project (a slight amount of paraphrasing has been done):

(1) What the product is, or what you want it to be: a main-dish product, a meal replacement, a standardized food, an imitation of a standardized food, a food subject to nutritional quality guidelines, or a food for which a common or usual name has or has not been established?

(2) What ingredients are you putting into the product? Are there additives such as nutrients, imitations of natural products, artificial flavors, etc.?

(3) The amounts of certain ingredients having special treatment under regulations. What needs to be added by the consumer? Percentages of vitamins, minerals, protein, fat, etc.

(4) What quantity will be sold in a marketing unit? What is the reasonable serving/ portion weight? What data do you have to support your position in the absence of a specific regulation? How many units in a package?

(5) How do you want the consumer to see the product? Will you show a illustration of fresh fruit, when the product is artificially flavored? Do you want to show a dish of ice cream, when your frozen dessert does not meet the standard of identity for ice cream?

Baumann (1975) devised an extensive check list of "the various steps toward legality" in designing new products; it is concerned primarily with improving or simplifying the interactions between legal, R&D, and marketing functions. The list is too extensive to include here, and also is out of date in several respects. Readers interested in this subject may consult the Bibliography for the location of Baumann's article.

Technical personnel should use great caution in contacting Regulatory Agencies. Complete transcripts of the conversations and complete files of all correspondence involved in these contacts must be kept, essentially, forever. On the other hand, discussions between company employees on sensitive topics are best kept in oral form (insofar as that is possible) since memos and other documents may be subject to misinterpretation (always to your disadvantage) if they must be delivered to an opposing attorney or to a court in civil or criminal actions.

If a representative of a regulatory body visits the plant, either on a routine inspection or in response to some sort of complaint, or perhaps only to fill his or her quota for the month, contact should be restricted to as few people as possible, only one person if that can be arranged. In any case, an observer should be delegated to dog the visitor's steps and record his comments and actions.

For day-to-day evaluation of the legal questions concerning proposed formulas, processing methods, ingredient compliance package structures, graphics, distribution methods, advertising campaigns, etc., it is desirable to assign one member of the product development team to keep current with the regulations as they are issued or proposed by agencies of the federal government and of localities where the product will processed, stored, or sold. This "expert" cannot be expected to replace the attorney's judgement, but will supplement the attorney's advice and assist him in understanding the technical problems.

Sources of information that can be consulted by the in-house regulatory contact are: the **Federal Register** (complete, but often difficult to interpret), advisory services such as **Food Chemical News** (valuable for interpretation, predictions, analyses, etc.), the news media (generally late, inaccurate, and biased), technical and semi-technical publications such as **Food Engineering** and **Prepared Foods** (often very good insightful commentary but sometimes delayed by printing schedules and incomplete due to editorial exigencies), trade organizations such as Grocery Manufacturers of American (can be helpful if you are a member, especially if you are a very heavy contributor), and advice from suppliers (quality of information ranges from very good to very bad and is frequently strongly biased).

It is almost always bad policy to ask a Federal agency (or a state employee) for an advance decision as to the legality of a food, container, or label. After incredible delays, you will either get a definite "No" or, more likely, an evasive answer that leaves you worse off than you were to begin with. Few bureaucrats are willing to smooth the way for you, because there is nothing in it for them.

STRATEGY FOR MANAGEMENT MEETINGS

Interfacing becomes a major problem when the technologist is required to participate in meetings that include several members of top management. Such meetings can occur several times during the course of a new product project. At least one of the meetings – and a very critical one – will take place when final approval is being sought for introduction of the new product into the market place. At this point, several reputations may be on the line, and there is an understandable reluctance on the part of many participants to take the gamble.

The product development manager should take a positive attitude, deprecatory statements and humility are definitely not part of the agenda. Do not dwell on difficulties that have been solved, any mention of problems will stick in the mind of the committee members long after the solutions have been forgotten.

Technicians are likely to be less effective than other executives in meetings of department managers and the like because of their tendency to rely on facts rather than feelings. Some diplomacy is required, facts are not always the final answer.

As so cogently expressed by Kelly (1990), "When engineers get together, chances are they gather round a calculator and some drawings in an urgent search for truth. There may be differences of technical opinion, and tempers may soar, but nearly all engineers are anxious to find an optimum solution, whatever that may be, and rally support for it. A managerial meeting, on the other hand, is more likely to be an adversary proceeding – not an open clash of wills, heaven forbid, but still a set of opposing viewpoints trying to obtain dominance with very little concern for truth and beauty. Such a meeting will be conducted in a genial, hearty, Dale Carnegie atmosphere, but in essence it is a game of getting the upper hand. If you would survive in this environment, it is important that you learn how the people interrelate with each other, what roles, they play, and what weapons they consider acceptable."

I am not sure that all engineers/technologists are the high-minded, non-manipulative persons implied by the preceding quotation, but the general direction of Kelly's comments seems accurate, based on my personal participation in hundreds of management meetings. What, then, is the technologist to do when participation in meetings of this type becomes necessary? The basic approach is to be well fortified with objective information to support your viewpoints, try to foresee the inevitable complaints and harebrained suggestions (and prepare replies based on logic and evidence), and determine never to let yourself be coerced into acquiescence with a goal that is either illegal or highly improbable of achievement. Never lose your temper, that gains nothing.

In extreme situations, we may observe behavior similar to that reported by Tompkins (1992), who describes an actual management meeting for new product evaluation in which the following occurred: "There had been no less than ten creative suggestions on how to reduce production lead times and the production manager explained why each of these suggestions would not work . . . Then, to make certain that the last ounce of creativity was driven from everyone's mind, the production manager concluded the meeting by stating that 'these meetings, where individuals shoot from the hip, are a waste of time and from now on no one will make suggestions until they have fully evaluated the practicality and economics of their suggestions.' This bully of a production manager had: 1. Made 100% certain that no one would offer suggestions for improvement. 2. A full understanding of the present operations and why anything suggested would not work."

BIBLIOGRAPHY

BAUMANN, H. 1975. A step at a time builds new products. Canner/Packer *144*, No. 12, 20-21.

BROWN, J. K., and KAY, L. W. 1988. Tough challenges for R&D Management. The Conference Board, NYC

DOPP, M. 1993. The compliance officer is coming. Food Processing *54*, No. 5, 18, 22-23, 26

FRAND, E. A. 1984. Marketing-laboratory interaction. Res. & Dev. , No. 12, 23.

GELLER, J. H. 1977. The regulations affecting product development — government view. Food Technology *31*, No. 7, 78-79.

GOLDENFIELD, I. 1977. The regulations affecting product development — industry view. Food Technology *31*, No. 7, 80-82.

HEGENBART, S. 1990. The R&D and marketing melee. Prepared Foods *159*, No. 3, 117-118

KELLY, J. 1990. Surviving the management meeting. Production Engineering *24*, No. 5, 24-27.

LASTER, R. 1976. Strengthening technology's role in product success with better communication at the marketing/R&D interface. Production Engineering *10*, No. 4, 20, 22.

MONTELEONE, J. P. 1976. How R&D and Marketing can work together. Research Management, March April, 19-21.

MULLICH, J. 1993. R&D vs. marketing: Time for a truce. Food Business *6*, No. 13, 9-12.

PEARCE, S. J. 1987. Quality assurance involvement in new product design. Food Technology *41*, No. 4, 104, 109-110, 113-114, 116.

REISS, E. S. 1977. Advertising and product development. Food Technology *31*, No. 7, 75-77.

SCHENGRUND, D., and MOELLER, D. R. 1980. Successful product development: marketing, research perspectives. Candy & Snack Industry *145*, No. 4, 37-38, 40.

STRUSE, R. W. 1977. Marketing and researching new products. Chicago Marketing Scene, Spring issue, 1977.

SWIENTEK, B. 1993. Hershey "Hugs" team building. Prepared Foods *162*, No. 11, 44-45, 48, 50.

SZMEDRA, P., and FERGUSON, W. 1991. The Delaney clause: new interpretations. Food Review *14*, No. 2, 6-7.

TOMPKINS, J. A. 1992. Intelligent ignorance. Industrial Product Bull. *49*, No. 8, 39.

WESTWOOD, A. R. C. 1984. R&D linkages in a multi-industry corporation. Research Management, May-June, 23-26.

WESTWOOD, A. R. C. 1988. The R&D-company interface. *In* Tough Challenges for R&D Management, J. K. Brown and L. W. Kay, Editors. The Conference Board, NYC

chapter eight

CONTAINER DEVELOPMENT

INTRODUCTION

Although ideas for novel containers may originate prior to the identification of a specific food material to be packed in the container, it is far more common for product ideas to be concerned with a new foodstuff, the package being designed to meet the requirements of its contents. At least, this is the case with R&D efforts in the food industry; development teams working for manufacturers of packaging equipment and packaging materials may have a different point of view. As an example of the "design the package first" contingent we have Primo Angeli (as quoted by Pierson 1993), who claims, "The earlier a marketer can discover a package that has the potential for capturing the consumer's interest, the faster and more effectively he can distill the essential ingredients necessary to develop the product that will sell itself right off the shelf." But, on the other hand, most consumers want a food that tastes good and provides good nutrition, characteristics the majority of containers lack.

Meanings of the words "label," "package," "product," and "container" differ somewhat among authors. For the purposes of this book, "container" will be defined as the structure immediately adjacent to the product and serving to confine and protect the product. For a can of peas it would be the metal can with its protective inner lining (if any) and any sealing compound present in the rolled edges, but not the label pasted on the outside; for a sack of sugar it would be the sheet of paper or plastic laminate used to form the bag together with the thread and associated material used to close the formed and filled bag; for a loaf of bread it would be the bag formed (usually) from a plastic laminate and the metal-wire or plastic tag used to hold the twisted end together.

"Package" will mean the container plus all materials outside the container except those used specifically to convey information from the manufacturer to the user or as required by government's edicts. The package is expected to help sell the product, to assist in attaining an shelf-life adequate for the system to be used for its distribution, to give at least some protection against mechanical damage during transportation and handling, to provide a surface on which to print expiry dates and legally required information, and to be convenient for the final customer to open and reclose. It is the task of the package designer to achieve all of these goals and more, in a package that can be made at the minimum possible cost in materials, equipment, and labor. "Label" will include printed pieces on or in the package that are added to the package specifically to convey information (including advertising copy) in the form of text or graphics.

"Case" is an outer structure, often made of corrugated board, designed to surround and protect one or (usually) more complete packages of the consumer unit of sale. One dozen is a common number of consumer units contained in a case, but this amount should be considered only a starting point, if that, in package development (Friedman 1975). Originally, casing in groups of twelve had some merit as it allowed an internal arrangement of 4X3 units that often led to economies in case materials and, of course, dozens are traditional ordering units for grocers. Intermediate groupings, such as six packs of beer cans, cartons of ten cigarette packages, etc., that are designed to provide convenience for both consumers and retailers, are probably best regarded as hybrids of labels and cases.

Development of the container for a new product should take place at the same time as development of the foodstuff, if that is at all possible. The most obvious reason for this approach is that timeliness of product introduction is improved, since the whole duration of package development does not have to be added on to the time required for food development. Also, the goals and specifications guiding development of the foodstuff may have to be altered partway in the project because of of some unexpected interaction of the food and package. In many programs, it will be found to be impossible to develop a satisfactory food product independently of its package.

Development procedures for containers will be covered in this chapter, labeling in the next, and regulations in the chapter following that.

WHAT DOES THE CONSUMER WANT?

It seems almost axiomatic that attractive packaging of foodstuffs in containers that adequately protect the product is necessary for successful marketing. Appealing to reverse snobbery is sometimes effective, however; limited success has been obtained with cookies marketed in a printed kraft bag and with French-style baguettes placed in a paper sleeve. Even these specialties would be acceptable only for retailers having very short distribution channels and requiring just a few days shelf-life — they would pose many difficulties for the large-scale manufacturer distributing products on a nation-wide basis. Of course, generic foods sold almost entirely on their price-appeal are expected to be packaged in the simplest type of container.

Several studies reporting consumer attitudes toward existing forms of packaging have been published. A consumer "want list" is said to indicate preferences for packages that are reclosable, storable, heatable, resealable, transportable, disposable, and biodegradable (Anon. 1988A).

Copeland (1988) compiled a check list of what consumers want in packaging for foods in general, not just bakery items. This included: (1) Packages that are lightweight, tamper-evident, and (when appropriate) microwaveable; (2) Aluminum cans that are resealable; (3) Packaging that extends life and improves quality for fresh produce; (4) A single address to write to and a single telephone number to call for information or for registering a complaint; (5) Colors that help delineate lettering – as a bad example, printing red on yellow makes reading difficult; (6) Packaging that looks good but isn't good enough to save (so the consumer doesn't feel guilty about throwing it away); (7) Directions the entire family can follow; (8) Freshness dates and expiry dates; (9) Nutrient labeling that is meaningful, easy to read, and has relevant comparisons; (10) Accurate and simple directions and graphics; (11) Packages that are truly easy-opening; and (12) Packagers who recognize their responsibilities regarding littering, non-biodegradable material, and the environment. Other features most retail buyers would appreciate might include: (1) Package dimensions and weight that make for ease of handling and shelf-storage; (2) A design that provides a convenient method of dispensing bulk ingredients, breakfast cereals, and the like; and (3) Ease of disposal.

Some of the items on this wish list will be difficult to achieve and of limited marketing value – no doubt many consumers that say they want such things as ecologically-friendly containers will, when confronted with a choice, buy the unopenable, nondestructable, and poorly labeled package with a cartoon character on the front and a contest coupon on the back.

SOME BASIC CONSIDERATIONS

If the R&D team includes a packaging technologist, completion can often be greatly expedited and mistakes prevented. However, small companies may have food technologists who must double as packaging specialists – one of the few advantages of this arrangement is that it ensures integration of food and container development effort. In any case, the packaging development scientist should be made part of the overall team and should be under technical direction (not a part of Marketing, for example). Companies who delegate the packaging development function to non-technical departments such as purchasing or marketing exhibit a poor understanding of the true nature of the development process and may not only experience costly errors but fail to achieve any creative results.

The structuring of a packaging development project will, in general, follow the same pattern as that of a new foodstuff. Schier (1974) recommends establishing a "Packaging Profile" at the start of a project. This outline of objectives (1) describes the new product, including its food chemistry, (2) includes organizational marketing objectives, (3) details protective requirements, (4) sets company merchandising goals, (5) estimates package costs on the basis of profit structure, (6) establishes graphic and label requirements, and (7) schedules each step of the development.

Schier suggests that the system be initiated and followed by a coordinator who is competent in establishing packaging criteria. The coordinator consults with all the departments involved and with outside suppliers before submitting the packaging profile to management for approval. The profile should be initiated at the time of the new product's conception, not after its development is completed. This enables the packaging coordinator to develop prototypes in various materials and forms. As time passes, the profile is updated, taking advantage of knowledge acquired in the interim. Changes should be specified in written form and circulated among personnel who will be affected by them.

Anthony (1982) describes a system consisting essentially of five steps:

(1) Gather information efficiently, as by:

(a) Formally establishing lines of communication.

(b) Developing well thought out checklists that provide structure and organization to information gathering.

(c) Starting the information-gathering process early.

(d) Developing strategies for sharing information having mutual benefit and interest.

(2) Formulate requirements, keeping in mind there are low to high levels for each functioning, with corresponding costs.

(a) Express requirements in the form of specifications and standards.

(b) State packaging requirements in measurable terms, as by specifying standard tests.

(c) Allow the providers of the information on which requirements are based to to review these requirements.

(3) Choose the materials and design the package. Design for simplicity. Be aware of the pitfalls of innovations.

(4) Test the packaging, using the following guidelines.

(a) Conditions should be identified and quantified and should correlate to field conditions.

(b) The most important requirements should always be tested.

(c) Tests should be performed according to written procedures.

(d) Abandon or revise a test procedure as soon as it is discovered that the test is not predictive.

(e) Stay sensitive to costs.

(5) Improve the packaging. Be willing to make changes in response to inputs of information on requirements, materials, processes, regulations, consumer preferences, etc.

ORGANIZING THE PACKAGING DEVELOPMENT FUNCTION

Location of the packaging development group in the table of organization varies between companies. Marketing views the graphics connection of packaging as the major function, and often works hard to get the group placed as a subdivision within their department. Production and/or Engineering regard the machinery and materials aspects of package design as of primary importance, and strive to get control of the group, regarding themselves as the primary custodians of engineering know-how. R&D sees the technical facets (chemical, biological, and physical) as the basic principles involved in container development, and also perceives the close relationship that must exist between food and package design during the entire product development phase, and so recommends Package Development be placed under R&D management. The CEO may want this important function to remain under his direct control, which means (in large companies at least) that control will be intermittent, casual, and lacking in scientific content.

There is no clear-cut solution to this problem, but one thing is certain: the various operations involved in package development should not be separated — a fragmented group will not function efficiently. In the author's opinion, the three best locations for the package development group, in order of decreasing desirability are: (1) R&D, (2) Engineering, provided it is a department separate from Production, and (3) A separate department reporting directly to the CEO. The three worst arrangements, in order of increasing undesirability are: (1) Production, (2) Marketing, and (3) Purchasing.

An article by Fernandes (1978) contains suggestions for organizing package development functions that are interesting and perhaps useful even though his paper is a bit outdated and is concerned principally with drug packaging. He lists the following objectives for the development process:

(1) Protect the product contained.
(2) Best serve the needs of the end user.
(3) Enhance product appeal through the use of appropriate design and graphics.
(4) Achieve optimum efficiency and economy in manufacture.
(5) Keep component costs to a minimum consistent with individual product design and quality objectives.
(6) Conform to all applicable regulations.

Fernandes points out that, in order to accomplish these objectives, the package development section must coordinate the participation of several departments and groups in four critical areas: (1) Container selection and specification; (2) Selection and engineering of packaging materials; (3) The design and application of graphics; and (4) Packaging equipment and utilization. Key departments are research and development, marketing, purchasing, quality assurance, production, and management, as well as non-company participants such as advertising agencies and market research consultants.

FUNCTIONS OF CONTAINERS

There are many considerations affecting container quality and suitaility for a given purpose that cannot be discussed here because of space limitations or because they do not conform to the technical orientation of this book (i.e., they are more often the responsibility of Marketing, Purchasing, etc.). It is important, however, to discuss interactions of the container with the product and with the machinery used to apply the packaging material, since these are facets of container design for which the development technologist must take responsibility.

Container development must take into account the shape and dimensions of the product-confining structure, the materials used to make it, the treatments applied to it, and the packaging equipment used to form, decorate, fill, and seal it. Interaction of the container with the food it encloses must be considered. Most food technologists have a rather limited knowledge of these specialized branches of technology and engineering; this is one of the reasons it is wise to include a packaging engineer in the development team.

Protection of Contents

The package must protect its contents from gross contaminants that are constantly or intermittently present in its environment. Dust, dirt, water droplets, insects, loose or smeary material from other products, and the hands of consumers or employees must be kept away from the product to the greatest extent possible within the constraints of cost and labor under which the manufacturer operates. To perform these functions, a hermetic seal and an impervious material are not absolutely necessary, but a tamper-proof seal and complete coverage by a moisture-resistant film or board are highly desirable. Packages meeting only these minimum requirements might be satisfactory for in-store bakeries, pizzerias, and other operations making products that will be sold in proximity to the production area within a few hours of completion.

Protection against mechanical damage. The package should protect its contents from damaging physical forces, as from crushing, cracking (e.g., must offer a non-flexible platform for pies), and punctures (as from fingers or corners of adjacent cartons). Inverting of the package can cause messy results with iced cakes, cheesecakes, pies, and most fragile items, but it is difficult if not impossible to design a container that gives good protection against such occurrences. A moderate amount of protection to fragile foods in pouches (potato chips, popped corn, etc.) can be obtained by adding enough air or other gas to the package so that the pouch becomes a puffy "pillow" within which the pieces move loosely. Pressure on the container is then resisted by the gas-supported film instead of the product. Such packages give much greater freedom of movement to the chips, etc. (as compared to tight pouches), possibly allowing greater damage from abrasion. And, added film is required, so packaging costs increase.

It would be desirable to be able to protect delicate products from damage caused by vibration, which can have disastrous effects on most types of pies, many cakes, delicate pastries, etc. As a practical matter, though, packages giving protection against gross abuse of this type are not available, and vibration damage can be prevented only by providing suitable transportation means and by training employees in preventative techniques.

Consumers do not expect all products to be protected equally. Obviously, a polyethylene bread bag gives only minimal protection against intrusive forces but it is well accepted by consumers, partly because it gives them an opportunity to judge "freshness" by squeezing the loaf. Breads, buns, and the like are usually protected in transit from bakeries to retail stores by baskets, trays, and other multi-pack carriers that prevent crushing. Foods generally acquire much more mechanical strength and resistance to damage when they are frozen, and this greatly facilitates nation-wide distribution of frozen cream pies and the like.

Protection against gas transfer. Most foods require protection against moisture-vapor transfer after they have been placed in their package. If this is not done, crisp products (such as crackers, breakfast cereals, and snack chips) can quickly absorb water vapor from the atmosphere and become soft or even soggy, while soft products (such as bread) can become tough and undesirably firm. As a first approximation, it is fair to assume that the product when filled into its package is at a moisture content considered to be optimal for consumer acceptance. Any gain or loss of moisture will then lead to a reduction in the organoleptic value of the item.

Interchange of moisture between the product and the atmosphere surrounding the package can be prevented only if two conditions exist: (1) The container must be constructed of a material that has a very low moisture-vapor transmission rate, and (2) The package must be hermetically sealed with no openings that would allow moisture vapor to flow in or out of the package. The common metal can with sanitary seams is an ideal container from this standpoint, but it is, of course, impractical for many products sold through normal commercial channels. It is also relatively expensive. A well-sealed pouch made of a film containing a layer of aluminum foil is the next best choice. Some products are not too demanding in this respect because their shelf-life is expected to be short — fresh unfrozen bread is satisfactorily protected from drying by the common polyethylene or polypropylene bag for the few days it must last before it is consumed.

Many foods are very susceptible to flavor damage resulting from absorption of aromatic substances present in the surrounding atmosphere. Vapors originating from gasoline, tobacco, insecticides, toiletries, and the like are very harmful to mild flavors and delicately balanced aromas, such as those characterizing bakery products, confections, etc. Some of these odorous compounds cannot penetrate the plastic films commonly used to package bread and similar foods, but petroleum-derived substances and certain other non-polar compounds will readily penetrate polyolefins, cellulose films, paperboard, and certain other well-known packaging materials. The good resistance to moisture vapor transmission exhibited by plastic films is not necessarily associated with a high level of protection against odor transfer, but aluminum foil layers will effectively prevent any translocation of aroma as well as water.

The package should also prevent transfer of desirable flavors from the product to the environment. Flavor may be absorbed into the container material itself, a phenomenon sometimes called "scalping." Freezing may delay but will not necessarily completely prevent flavor loss by these mechanisms. Since flavors are often very complex substances, usually consisting of delicately balanced combinations of chemical compounds having very different physical qualities, it is difficult to predict which type of container material will be the most effective in preventing flavor loss. Paperboard and polyolefin films are not very resistant to the transfer of odors; plastics other than polyethylene and polypropylene may be more helpful. A layer of foil will certainly prevent odor transfer into and out of the container, but it will not prevent absorption of desirable odorous materials by plastic layers located inwardly of the foil.

Attempts have been made to solve the scalping problem. For example, a process has been described that treats the inner surface of a blow-molded polyethylene container with fluorine to form a very thin continuous layer said to act as an excellent flavor barrier. The surface is modified to a depth of only about 200 to 1,000 Angstroms. The FDA status of this treatment has apparently not been resolved, and it is not known whether or not it has been used for commercially produced food products.

It will often be found that it is necessary to prevent movement of oxygen molecules across the package barrier. Other gases are occasionally troublesome, but oxygen usually creates the most problems. It is particularly important to keep out this gas when anaerobic conditions are being established by flushing the package with nitrogen. Proper selection of the packaging film will allow maintenance of the reduced oxygen level for quite a long time, provided the seals are adequate.

If an anaerobic atmosphere is desired for controlling fungi, nitrogen is usually the gas chosen because it is cheaper than, say, carbon dioxide, and it does not have some of the side effects of the latter gas. For effervescent soft drinks, it becomes important to retain carbon dioxide inside the bottle, almost always under considerable positive pressure, and this requires thick layers of specific barrier materials. A plastic showing good protection against nitrogen transfer does not necessarily have a low transmission rate for carbon dioxide.

Protection against radiant energy. Products requiring protection from light and ultraviolet rays must be packed in containers that are substantially opaque, such as those made from paperboard or from metallized films or films containing an aluminium foil layer. An important negative feature of such packaging is that the product cannot be seen by the prospective buyer. Products being marketed through fresh distribution channels (many bakery products fall into this category) ordinarily do not require this protection, because their shelf-life is so short, but opaque containers may be necessary for foods susceptible to the development of oxidative rancidity during their normal storage period. Light or ultraviolet radiation tends to accelerate development of oxidative rancidity.

Actinic changes in surface pigments, as may occur in icings and coating colored with a mixture of dyes, can lead to bizarre and very unappetizing hues after moderate periods of exposure to sunlight or other intense sources of actinic rays; these changes result from different rates of of deterioration of different dyes. The brown color in the crusts of bakery products may be lightened by similar processes. Makinde et al. (1976) pointed out that some vitamins (e.g., riboflavin and vitamin C) can be destroyed by light-mediated interaction between the vitamin and oxygen.

Occasionally, the manufacturer will want to provide some insulation as a protection against rapid temperature changes. No thin film will serve this purpose. Foamed polystyrene does provide some insulating properties, but to be effective it must be used in thicknesses that are awkward for many applications and essentially impossible in some. It is also difficult to use and has little mechanical strength. Paperboard has some value for insulating purposes, but different grades vary greatly in their effectiveness. The "puffy pouches" previously described as a means for reducing crushing may also slightly reduce rates of heat transfer, but there is no really satisfactory solution to the insulating problem in film packaging.

Display Requirements

When a package is being designed, the Marketing Department will want to know if the package surface will be compatible with the graphics it deems desirable for selling the product. Among the questions they will want answered are: (1) What kind of printing methods (flexographic, rotogravure, letterpress, etc.) are acceptable? (2) What type of inks can be used? (3) What is the unprinted color of the film, and can this be modified if necessary? and (4) Is the film opaque, translucent, or transparent?

There must be sufficient space available for product identification, selling messages, notices required by laws and regulations, bar codes, and other more or less essential words and images. Opening and reclosure features must be acceptable. Except for some unusual applications, the package must be constructed so as to stand upright on the supermarket shelf, providing maximum visibility to the consumer and convenience of grasping when being removed from the shelf.

Some of these display requirements may lead to the need for an overpack to contain one or more unit packages. The outer carton will seldom be required to afford all the protective features demanded in the interior packages although it may, in fact, be cheaper to transfer these functions to the outer container when it contains several unit packs. In the latter situation, the interior packets nay be made of some cheap material such as thin polyethylene or coated paper. This kind of packaging is acceptable only if it is reasonably certain the interior packets will be used up before uptake of moisture, oxygen, or other harmful substances renders their contents unacceptable.

Convenience features for the consumer

There are many convenience features that the packaging technologist will want to consider introducing into the container design during the course of a development project. Some of those which might be useful are related to opening features, portion packs (including vend packs), bake-in containers, and adaptability of the container to microwave heating.

Opening features The consumer must be able to open the package easily without using specialized utensils and without having to read complicated directions. The contents of the package must not spill or soil the hands during the opening process. The product must not be damaged by the opening procedure. If the package is to be reclosed to protect the food that remains after one or more portions have been withdrawn, the reclosure feature must be simple and reliable and it should preferably be an integral part of the package so that it does not become lost, fall down the garbage disposal, or otherwise infuriate the consumer. In several market research studies made to determine what consumers dislike most about modern packages, difficulty in opening the container ranked very high on the list.

Portion packs. Individual serving packets are often used as samples in preliminary consumer acceptance tests of food items. Although the final package concept may not be testable under these conditions, the food idea should be susceptible to evaluation. Much money can be saved, especially in large tests involving foods that are difficult to produce in small quantities. For the product developer, the relevance of this observation is that it points a way of determining consumer reaction to a new item at reduced cost for test material.

Portion packs can often facilitate entry into markets that, though limited in absolute numbers, offer an opportunity to appeal to customers not now being served. Some suggestions for portion packs of particular applicability to the food and beverage businesses are: packaging variety breads in single serving (e.g., two 1-oz slices) pouches, multipacks of single servings of drink powder of the Kool-Aid type, packets of single servings of microwaveable sauces and gravies, and boxes of perhaps 8 to 10 soda crackers. Such miniature packs would be of special value for dietetic products, meeting the needs of households in which only one or two persons require these special foods.

The development technologist may find it necessary to design a vending machine package to meet a Marketing Department plan for penetrating this demand segment. Preliminary guidance should be obtained by studying the practices of other industries serving this market. Compliance to dispensing machine limitations and the special distribution system prevalent in the industry servicing these machines will be essential. Assistance, or at least advice, can probably be obtained from manufacturers of vending machines and from the purchasing departments of companies selling goods through these devices. The packaging technologist must determine, and rigidly comply with, dimensional limitations and storage life specifications.

Bake-in containers. It may be a marketing advantage to allow consumers to use the package as a baking container, as for finishing off bread and roll doughs, pies, cakes, and the like. For many years, brown 'n serve rolls have been distributed in foil and paperboard pans suitable for this use. Aluminum foil is a good choice for baking in conventional ovens although it becomes expensive in the gauges needed to give adequate handling poperties. The major disadvantage of foil is that it is unsuitable for use in the microwave oven. Ovenable paperboard and ovenable plastic pans have been developed to permit the baking of foods from the frozen state, and have had considerable success in that application, though cost becomes a serious problem here, too. The potential of heat-resistant plastic for the home cooking of bakery foods and certain other types of foods does not appear to have been fully exploited; there are many possible uses that have received very little attention from manufacturers.

If the package, or some part of it, is to be used as a baking container, consideration should be given to designing pans that can act as serving dishes. This may involve merely giving greater attention to esthetics, though some technical features such as greater rigidity and resistance to puncturing and cutting may also have to be considered.

Adaptability for microwave heating. In recent years, there has been a great proliferation of foods (usually frozen foods) packed in containers that enable the product to be heated in a microwave oven. Many different types of foodstuffs and combinations of foods in the form of a meal are being sold. At the start of this evolution, most of the products being offered were fully cooked foods that required thawing and warming in the microwave oven. These initial efforts were quickly followed by products needing both thawing and cooking, and therefore demanding containers having much more resistance to heat damage.

Bakery foods particularly suitable for thawing and warming in microwave ovens include fully baked, single component (i.e., no icing or filling) structures such as pancakes, soda biscuits, rolls of various sorts (croissants, etc.), pre-baked fruit pies, and many kinds of cake-like and cookie type items. Full-size loaves of bread do not respond very well to this type of heating because energy penetration is too erratic, but muffins, buns, and small loaves of the quick-bread t, pe are being offered. If the product is to undergo a baking process in the microwave oven, it must be of a kind that does not require crust browning since uniform coloration is virtually impossible to achieve in this equipment. Even those ovens with a resistance heating element in the top of the cooking chamber do not brown the surface of products as well as could be desired.

Doughs and batters designed for baking in microwave ovens pose more difficult packaging problems than thaw-and-warm products. From the convenience standpoint, it should not be necessary for the consumer to transfer unbaked material to a plastic or glass pan before placing it in the cooking equipment. Therefore, it is a marketing plus to pack the product in a container that can be placed directly into the oven. Metal pans are usually not suitable for this purpose; although some products have appeared in foil containers said to be adapted to microwave baking it is believed they have received only limited acceptance. The best containers for microwave cooking are non-metal trays or pans constructed from polyester-coated paperboard or thermoformed from certain types of rigid plastics. The container itself must neither absorb nor reflect significant amounts of the microwave radiation, it must not soften appreciably when the food in it becomes hot, and it must not give off toxic, odorous, or flavored materials under any set of circumstances it is likely to encounter.

Cake batters made from dry mixes can be baked reasonably well in the microwave oven. Of course, special formulations are required, only small pans are suitable, and crusts do not brown adquately, but these are not insuperable obstacles for cakes intended to be made and consumed within the home. The tops of the baked layers frequently have a wavy appearance resulting from uneven distribution of radiation in the oven chamber. Brownies respond well to microwave conditions, but cookies from individually deposited lumps of dough usually do not.

Polyethylene terephthalate (PET) is a plastic resin that has been recommended for use in trays heatable in either microwave or conventional ovens (Anon. 1988C). The manufacturer's literature indicates that a PET sheet can be thermoformed into ovenable trays provided a nucleating agent is included in the resin to promote fast crystallization during the molding process. Without this additive, thermoforming is not practical.

The shape of the container has a significant effect on its performance in the microwave oven, especially with regard to uniformity of the cooking process. Rectangular containers should be avoided, because they contribute to uneven heating — although rectangular trays with rounded corners have proven to be satisfactory in some situations. Round or oval trays are preferred. The height of the bottom of the tray from the floor of the baking chamber also has an effect on uniformity of bake. Heat distribution is made more even by covering the product with a transparent film such as PVDC; this holds the hot vapors in contact with the surface of the food. Placing in the oven chamber a revolving platform on which the baking pan is placed is a considerable help in obtaining uniform baking.

Special pan designs include the addition of a layer called a "susceptor." It absorbs microwave energy, turns it into thermal energy, and causes localized heating. When placed under the bottom of a pizza, for example, it can cause browning and crisping of the lower surface without overcooking the top of the product. This helps to prevent the soggy, soft, and generally unpleasant crusts that are so predominant in microwave-cooked frozen pizzas. The active component of most susceptors is a layer of powdered aluminum laminated under plastic films.

Legal Considerations

The laws and regulations governing packaging of foodstuffs are becoming increasingly restrictive. All package designers must keep abreast of legal developments if they are to function effectively. It is no longer safe to rely on supplier advice as to the legality of a particular food and package combination, especially if the supplier is accustomed to dealing mostly with nonfood manufacturers. Salesmen's comments such as, "This film is GRAS" or "This material is in the process of being approved by the FDA" are not an adequate basis for using a packaging material in the container for a new product. After much time has been spent in designing a package around the optimistic salesman's material, it may be discovered that he based his recommendation on the off-hand comment of some nontechnical person in his own organization or in a competitive company, or even on general trade gossip having no firm foundation whatsoever. Then, the technologist may be faced with a complete redesign of the package at a late stage in the development process, possibly with no satisfactory replacement for a package raw material that is still in regulatory limbo.

It is clear that food-contacting package components must be composed only of materials that have FDA approval, since some contamination of the foodstuff by the packaging substance is almost certain to occur during normal packaging and distribution operations. If the legal and technical staffs of the packager are not adequate to interpret applicable regulations, then the supplier may have to be relied upon to some extent. This means the user must, as a minimum, obtain unequivocal guarantees from the supplier that the specific material being supplied conforms to all applicable federal and local regulations for food packaging.

In the present regulatory climate, packaging materials that appear to be completely acceptable may suddenly become unusable. Sources of supply may be cut off suddenly because the supplier's plant is found to be violating some safety or environmental regulation and can no longer manufacture the needed material. Packages regarded as safe and acceptable by virtue of a history of long usage during which no consumer harm has occurred may become legal pariahs because a politically influential fanatic (or the public relations flack of some competitor of the supplier) states the resin or other packaging material causes environmental damage, is ecologically unsound, or has the potential of causing sickness in one of a million persons who eat a pound or so of the packaging material every day. A class of packages may become the subject of "deposit laws" that make their use much more costly or inconvenient. Many of these actions are unpredictable by any logical process and can be dealt with only after they occur. Constant review of news sources that specialize in legal activity in the food field is essential.

MATERIALS

The package developer must have a good grasp of the basic properties of the materials available from commercial suppliers. As a start in acquiring this information, periodicals in the fields of plastics, paperboard, converting, printing, and packaging should be reviewed on a regular basis. A late edition of one of the "encyclopedias" issued yearly by some of the packaging magazines should be on the technologist's reference shelf. If a material seems to have promise for a present or future project, samples and specifications should be obtained. The summary presented below is intended only as the foundation on which the developer can build. Examples of packaged food products quoted here are largely concentrated in the bakery field because that is the field in which the author has the greatest experience.

Factors Governing the Selection of Materials

Since a single film may not be able to furnish all of the desired features, it is common practice to laminate together several different films, each of which provides a necessary characteristic missing in the other members of the structure. For example, it may be considered necessary to use one layer for its heat seal properties, another for its printing compatibilities, a third as an oxygen barrier, a fourth for its moisture barrier properties, and a fifth for grease resistance. It is not uncommon to find structures having as many as eight layers, not all of them different. Of course, the cost per pound and per square foot of the film rises as the number of layers increases.

Although aluminum foil contributes many of the protective features desired in food packages — such as opacity, good appearance, and virtually zero transmission rates for all common gases, water vapor, and odorous compounds — it does have some disadvantages. It is relatively expensive, it is practically impossible to economically recycle film containing foil, it prevents the product from being seen by the consumer, and it is not easily adapted to shrink wrapping or stretch applications.

The requirements for a packaging material that would be ideal for bakery products might be summarized as follows:

(1) The material should protect the product from harmful environmental influences. Some of the ambient conditions that can be detrimental to

bakery products are moisture, temperature, oxygen, and light. As a primary requirement applicable to most cases, the packaging structure should protect the food from drying out during its expected storage life. Not only must the material have a low rate of moisture vapor transmission in the form in which it will be used, but it must also be capable of forming seals that are at least as impervious as the intact film. If the film performs adequately against moisture vapor transmission, it will very likely exclude mold spores, dirt, dust, and other foreign particles, and it will give some protection against the absorption of off-odors from the environment. The thickness of plastic films is directly but not linearly related to the rate at which water vapor can diffuse through the film. That is, a test conducted under standard conditions may show that a certain amount of water vapor will diffuse each hour through 100 square inches of a 1 mil thick layer of a certain kind of polyethylene; if a sheet of the material having a thickness of 2 mils is similarly tested, the rate of moisture vapor passage will be much reduced but probably not by 50%. Furthermore, the thicker film will cost less per pound (more per square inch) but the cost will not increase by double the amount due to the better efficiency achieved in manufacturing the thicker film.

(2) The package should contribute to the dimensional stability of the product. Since most bakery foods are very susceptible to crushing, mechanical strength must be added to the container if the product is to survive storage, transportation, and merchandising without undergoing an unacceptable amount of distortion, breaking, compression, etc. If the product is sold in the fresh form, with delivery from the factory to the store shelf, and has a short shelf life (as do bread loaves), protective measures may take the form of rigid boxes, trays, or other containers that surround the loaves during their transfer from the bakery to the store and with which the consumer has no contact.

(3) The film should be suitable for applying easily and quickly by mechanical means, and preferably by machines that are already available. Fundamental requirements are that the material heat-seal readily and that the seals be firm and continuous. The film should be capable of undergoing rapid transfers and foldings by the wrapping device without cracking, tearing, or stretching.

(4) The package should assist in selling the product. For bread, this usually means that the product can be squeezed so the customer becomes convinced the bread is soft, and therefore fresh. In cases where visibility of the product is important, transparent films are imperative. A glossy surface definitely enhances consumer acceptance. Adaptability of the film surface to printing techniques is usually necessary.

(5) The film should be relatively low in price. It must have a favorable cost per square inch (a much more important measure than cost per pound) and the supplier's factory or warehouse must be located so as to make transportation costs relatively low. It must be capable of being stored under reason-

able conditions for a few weeks or months without undergoing significant deterioration in any of its properties (including machining response), and it must machine with a minor amount of waste.

Packaging films will exhibit different degrees of resistance to moisture vapor transfer, and will differ in their barrier properties relative to oxygen and other gases, light, etc. They also differ in cost according to their strucure and the market price for their constituents. Some of the factors affecting cost are the base resin used, the gauge or thickness of the film, coatings applied to the film, and other treatments besides coating applied to the film. Barrier properties may be changed in a desired direction by speciying base material (resin), thickness, and coatings.

Lamination of two or more films may be used to modify barrier, sealing, visual, and printability properties of the finished structure. Few, if any, film materials will be found possess all of the desired properties to an optimum degree. Shortcomings of one film may be compensated for by using combinations of materials. Thus, the lack of strength and rigidity in a transparent overwrap may be overcome, at least partially, by using bands or trays of tougher materials as supplementary structures. Laminating of two or more films, particularly of aluminum foil with resin films, is a way of taking advantage of the best qualities of each of the materials.

A survey of some of some packaging raw materials is given in the following sections.

Cellulosic Materials

The flexible films that have been used for bakery products include polyethylene, polypropylene, white bakery stock paper, waxed paper, cellophane, glassine, and combinations of these materials. Of course, other flexible materials have been used from time to time. A few decades ago, waxed paper was practically the only film used for bread packaging, and a small amount is still used for that purpose. It was first supplanted by cellophane (including coated varieties of this material), then polyethylene became widely used, and finally, polypropylene gained considerable popularity. Each of these films has some desirable features and some disadvantages. Cost factors rank high on the list of factors the bakery manager considers in choosing a packaging material, and machining efficiency is about as important as cost per pound.

White bakery stock paper. Bleached white sulfate paper (usually 28 lb basis) is used for bakery bags in many retail and a few wholesale operations, primarily to package crisp-crusted breads such as French or Italian loaves. In the wholesale trade, untreated paper may be used for gluing into bag form on bag-making machines. This paper has no heat-sealing properties. For retail operations, the material may have been treated with hot-melt wax by the papermaker so as to better retain freshness of the bread.

The coating is applied to one side, and penetrates the rather porous paper.

Bleached sulfite paper cannot be printed in bright and vivid colors due to its porosity. The chief advantage of bags made from this material are (for unwaxed paper) the rather low cost, the high "breathing" value (rapid rate of gas transfer), and the paper's wide and traditional acceptance for some types of bread, and for waxed paper are its low cost, whiteness, and fairly good quality retention characteristics.

Vegetable parchment. This material was developed to replace animal parchment, which is a thin, dried, treated animal membrane (often a layer taken from the inside surface of a cattle hide). It is made by immersing sheets of unsized paper in a bath of concentrated sulfuric acid, then washing and drying the material. Properties of parchment are dependent to a considerable extent on the characteristics of the fibers laid down in the base sheet, but generally include greater grease resistance and higher mechanical strength than found in the base sheet. The reacted paper can be plasticized, waxed, creped, or modified in other ways. It is useful as an extra bottom sheet in packages of bakery foods that are particularly greasy, but it is quite expensive per unit area.

Transparent cellulose films. When cellophane was first offered to bakers, there were only a few types available. In time, many more varieties with a wide range of properties came on the market. Film varieties are commonly identified by a combination of letter and number symbols that indicate some of the properties of the material. For example, all films having heat-sealable coatings have an "S" in their identification number, while "P" indicates a plain film without coating or heat-sealing properties; "M" denotes moistureproof qualities; "T" signifies a clear or transparent film while "C" is used for colored films; "A" indicates an anchored coating that can be used for frozen foods; and "K' is the symbol applied to cellophanes having vinylidene chloride coatings.

The so-called PT cellophanes (plain transparent) were never of any significant value to the baking industry because they would not heat seal and they were poor moisture vapor barriers, thus permitting rapid drying out of the foods. Specially modified coated cellophanes overcame these disadantages and have been widely used for bakery food packaging. Coated films with yields of 19,500 to 21,500 square inches per lb in 300 gauge thickness and of 14,500 square inches per lb in 450 gauge thickness have been espeially useful.

Vinylidene coatings have been developed, and they have generally proved to be of excellent value in that they increase the shelf life of foods by decreasing moisture loss. Originally, vinylidene-coated cellophane was available only in 450 gauge due to limitations of the coating process, but it later became available in a 300 gauge variety that was free of fisheyes and other nonuniformities.

Cellulose film has a grain direction resulting from the manufacturing process. It is weak across the grain and relatively strong in the other direction. In use, it is run through the wrapping machines so that maximum stress is exerted in the cross-grain direction. This procedure can provide a strong package. Weakness may result either from loss of moisture or pickup of moisture by the film with resultant shrinkage of the product and stretching of the film. On lengthy unsupported items, such as 12-, 13-, or 14-inch long loaves of bread, the moisture change effects can cause trouble in the form of tearing or bursting. In carton overwraps or in bags, few difficulties of this type are observed.

If the film is stored for long periods of times in spaces having low relative humidities, cellophane can lose part of the 5.7% to 6.0% moisture content it needs to remain reasonably limp and pliable. If such films are run in wrapping machines that have dull cutoff blades, fracturing of the film may become troublesome.

Cellulose films readily accept printing inks, and have an unexcelled transparency that adds to the attractiveness of the package's contents. They are inexpensive, at least under some market conditions, and are available in forms suitable for use in nearly all types of packaging equipment. They will seal within a fairly wide range of temperatures, producing a continuous and firm seal.

For most bakery packaging, cellophane has been supplanted by polyolefins – polyethylene and polypropylene. These polymers have much lower moisture vapor transmission rates and can be made almost as transparent as cellophane, although the "sparkle" is not there. Economics of the choice vary according to market conditions.

Paperboard. It is not unreasonable to consider paperboard, insofar as its protective characteristics are concerned, as being very similar to the flexible paper material described previously. Both are made from a suspension of wood pulp, which may be regarded as an impure preparation of chemically treated cellulose fibers. This pulp is deposited on a moving screen or other apparatus that allows most of the water to drain off and provides a support for the the sheet of paper or board until it dries. The amount of fiber deposited per given area determines whether the product is called paper or paperboard. The basis weight is expressed as pounds per 3,000 square feet, or (for some paperboards) pounds per 1,000 square feet. Outside of North America, basis weight (commonly called grammage) is expressed in grams per square meter. Paperboard has limited value in controlling moisture vapor transfer, but does provide some dimensional stability that may reduce damage to fragile products and it can provide good display area.

Special properties can be added to paperboard by subjecting it to various processes. For example, additives can be put into the pulp or applied to the wet sheet to add moisture resistance or change the color. The most common additive is an internal sizing ingredient, such as synthetic resins, that give the board some resistance to penetration by liquid water. A "setting agent," often aluminum sulfate, can be added to cause the resin to attach more firmly to the paper fibers.

Surface additives, such as grease-resisting materials, may also be added to the paperboard during its manufacture. The most common surface additive applied is starch, which seals the surface of the paper and provides increased surface strength, and more generally improves the printability of the board.

There are a number of grades of paperboard. Bleached packaging boards are often referred to as SBS (solid bleached sulfate) boards. These materials are used to make folding cartons and in some ovenable packaging containers. For folding cartons, a clay coating is applied to the side of the sheet that will be printed. A subsequent finishing operation (essentially an "ironing" or hot pressing step) provides the smooth surface desired for printing. Clay-coated news board is made of a different kind of pulp; it is cheaper, not as strong, and has a gray tint on the uncoated side. It can be used for some food cartons.

To be suitable for use in the fabrication of a folded box, paperboard must be able to withstand a 180° fold without cracking. Paperboard that meets this requirement is called "boxboard." Bending paperboard is generally classified as (1) boards produced from waste papers and (2) boards produced from virgin chemical pulps. Some of the many commercial types of the first class are bending newsboard, bending chip, solid kraft-lined chip, bogus kraft-lined chip, white patent coated news, white clay-coated news, bleached manila-lined news, manila-lined news or chip, and white vat-lined news or chip. These boards generally have only one coated side (or none), but if requested, both sides can be coated. White patent-coated and clay-coated boards are compatible with high-fidelity printing processes. Among the types of chemical pulp boards are bleached, semi-bleached, cylinder-produced, fourdrinier produced, coated, and uncoated.

Plastic films can be heat-sealed to paperboard to give it moisture- and grease-resistance, better appearance, special printing properties, sealing capabilities, and other features.

Plastic Resins

These materials are commonly divided into two major groups, thermosetting and thermoplastic. The former is primarily employed as the basis for re-usable plates, cups, and other dishes that do not have to withstand cooking temperatures, and for bottle caps and the like. Major varieties are phenolic, urea, and melamine resins. They will not be discussed further in this chapter because of their very limited use in commercial packaging of foods.

Thermoplastic resins soften and eventually become more or less flowable when they are heated. The melted materials become more viscous and stiffer when cooled, and set up into plastic or semi-solid masses below a temperature that is characteristic of each resin. The melting and solidification steps can be repeated, although eventually heat damage causes the resin to undergo irreversible changes in color, viscosity, etc.

At room temperature, different resins assume different degrees of rigidity, some having very little flexibility while others are soft and flexible. The cooling curve as well as the final temperature affect the mechanical properties of the finished object, since stiffness is related (in part) to the extent of "crystallinity" in the resin. The softening and stiffening cycle can be repeated many times (for some of the resins, at least) without causing significant changes in their physical or chemical properties. Prolonged or often-repeated heat treatment will, however, cause darkening and other unwanted changes in most commercial plastics.

Many of the thermoplastic resins approved for food contact purposes are used mainly in the form of films. Porner and White (1972) suggested that, to be successful, a film intended for packaging of foods must have the following characteristics: (1) Good moisture barrier properties, (2) Aroma barrier properties, (3) Grease resistance, (4) Barrier properties against absorption of off-odors, (5) Light resistance, (6) Acceptable coefficient of friction, (7) Good bond strength for coatings, thermal strips, and laminations, (8) Absence of blocking, (9) Good rigidity, (10) Good sealability, (11) Good appearance and feel, (12) Abrasion resistance, (13) Good machinability, (14) Good low-temperature durability, (15) Resistance to high temperature and humidity, (16) Toughness and tenacity, and (17) Protection of product freshness. No doubt there is some overlapping of these qualities, but it is a very comprehensive list.

Some of the most important members of the thermoplastic group of resins will be discussed in the following paragraphs. There may be limitations on the use of some of the compounds in contact with foods that are not mentioned in these brief descriptions.

Ethylene vinyl acetate. These resins are particularly suitable for the extrusion coating of paper, paperboard, and other substrates. EVA films have excellent impact strength, softness, and flexibility.

Ethylene vinyl alcohol. EVOH was first introduced in Japan in the late 1960s. These resins are hydrolyzed copolymers of vinyl acetate and ethylene. Their chief advantage is a gas barrier property significantly better than most other plastics. The gas barrier can be further enhanced by using a mica filler, although this additive makes the film significantly more brittle. EVOH films are being used as the barrier component in both flexible and rigid multilayer structures. Some applications are bottles and jars for ketchup, salad dressing, and jellies. Rigid retort trays and aseptic packages without foil layers are being thermoformed from multi-layered sheets containing EVOH. The resin is very sensitive to moisture, a feature making it unsuitable for the inner or outer layers of these packages; it must be "buried" inside moisture resistant strata if its gas barrier properties are to be fully effective. Use of EVOH in packaging foods seems to be relatively limited at this time.

Mylar. This plastic is a strong, tough, clear polyester resin made principally from polyethylene terephthalate. Its optical properties, dimensional stability, and durability combine to make it an ideal material for many packaging applications. It is available in both coated and uncoated forms. Uncoated film is used primarily as a substrate for laminating or for coating other materials. When used as an ovenable lidding film it provides heat seals that remain intact under wet, freezing, and microwave or conventional oven conditions. It promotes ink adhesion and bonding. Metallized coated and uncoated films provide visually attractive packaging and give excellent barrier properties. Mylar films coated on one side with a barrier material are used for packaging some kinds of bakery products, lunch meat, cheese, nuts, coffee, and other oxygen- and moisture-senstive products. Mylar films coated on both sides are readily heat-sealable and are used as a general purpose overwrap film and for lamination to other films.

Nylon. There are many different compounds sold under the tradename "Nylon." Some of these have been used as one or more of the layers in composite films used to package foods. For example, Nylon MXD6 (Harada 1988) was used in combination with polypropylene as the material for thermoforming containers that are to be sterilized in steam. This resin is said to have excellent gas barrier properties even after retorting. The compound is described as having high tensile and flexural strengths and low water absorption.

Polycarbonate. This polymer, which is produced from benzene derivatives, is a plastic of high clarity, very tough and rigid, and dimensionally stable. It is being used for five-gallon drinking water carboys, almost completely replacing the glass versions of these containers, and for nursing bottles. The cost is relatively high. It maintains its properties through a wide temperature range and has good resistance to many chemicals. It does not seem to be adaptable to the formation of films that are suitable for food packaging.

Polyesters. These condensation polymers are synthesized by reacting one or more dicarboxylic acids with one or more glycols. Polyethylene terephthalate (PET) is the common type, and it comes in varieties having widely different properties. PET resins are used as one of the component layers in most, if not all, plastic soft-drink bottles. This material is also used in blister packs for processed meats and as a coating for ovenable paperboard. When biaxially oriented, it provides a very good barrier against the transfer of odors and oils. It can be coated with other resins to make it heat-sealable or to improve its handling, barrier, or laminating properties.

As previously mentioned, PET is being used in dual ovenable trays (i.e., trays that can be used in either conventional or microwave ovens). For this application, it is critical to secure the proper degree of crystallinity in the resin before it is thermoformed into trays (Anon. 1988C). The aim is to produce trays having a balanced combination of low-temperature toughness and high-temperature dimensional stability.

Polyethylene. As the name indicates, this resin is a polymerized form of the gas ethylene. To make the plastic material, ethylene derived from petroleum raw materials is subjected to high pressure in the presence of activators and catalysts. The conditions of polymerization can be varied to obtain substances with a wide range of properties. The major classifications of importance to the packaging technologist are high density and low density polyethylene. Within these two categories are ranges of resins with varying characteristics, but, in general, high density polyethylene (sometimes called "linear") is stiffer, tougher, denser, and more opaque than the low density variety. The latter types are generally softer, more flexible, less crystalline, and more permeable to gases and moisture vapor. There are also intermediate grades, or medium density polyethylenes, that are made in much the same way as low density polyethylene, and the properties are somewhat similar.

Polyethylene is the polyolefin used in greatest quantity for packaging applications. It is made into films, sheets, coatings, molded containers such as bottles and jars, lids and other closures, and liners. LDPE is used in large amounts in making preformed bags for bread and the like; but it is not stiff enough in those thicknesses that are economically acceptable to run satisfactorily on most bread wrapping equipment. Medium density polyethylene films combine good heat sealability with good stiffness, while high density films offer superior stiffness but have poor heat sealability.

Cast polyethylene is the form in which most of this plastic is used for bread packaging. The term "cast" means that the molten resin is extruded through a very narrow slit and immediately cooled by passing it over chilled metal drums. This method produces films with good optical properties and stiffness. Oriented films are extruded by a different method by which they are stretched in either one or two directions as they solidify. This orientation of the molecular structure has very significant effects on many properties of the film.

Linear low density polyethylenes (LLDPE) are copolymers of ethylene and alpha olefins. These resins are composed of a mixture of molecules having differing degrees of branching, varying from highly branched to almost linear chains. The amount of co-monomer incorporated into the copolymer determines the number of chain branches while the length of these branches is dependent on the alpha olefin. Solid state properties of LLDPE are dependent on the length and amount of short chain branching, short chain branching distribution, average molecular weight, and molecular weight distribution. Further discussions on this subject can be found in the article by Parikh and Knight (1988).

The cost and availability of polyethylenes have been extremely variable over the rather brief history of this material. Most of this variability can be attributed to wide swings in the cost of the raw material – ethylene. Demand for both the raw material and the finished resin has increased remarkably, and sometimes exceeds supply (or apparently so). Long term predictions of price trends in any commodity are extremely risky, but it does appear that polyethylene will generally trend upward in cost over the next several years while remaining favorably priced relative to many alternative packaging materials. This price relationship is at least partly due to the fact that polyethylenes can be drawn into very thin films while many competing materials must be much thicker to function satisfactorily. As a rough guide, about 30,000 sq in of 1-mil film can be obtained from a pound of plastic.

Polypropylene. Polypropylene is a polyolefin, as is polyethylene, and the two substances have somewhat similar properties. The raw material or monomer, propylene, which is made from petroleum products, is a three-carbon chain as compared to the two-carbon chain of ethylene. Polymerization of the monomer is conducted in heated and pressurized reactors in the presence of catalysts consisting of alkyls and metallic salts. Polypropylene is considerably stiffer and much tougher than either high density or low density polyethylene of equivalent thicknesses. One very significant advantage of polypropylene is its superior machinability as compared to polyethylene.

Among the many varieties of polypropylene that have been offered to food packagers are cast oriented, cast-coated printed, oriented coated, and laminated combination films. Cast non-oriented film, the form in which polypropylene has been available for the longest time, is most commonly used in 0.97 and 1.00 mil thicknesses, but can be obtained in other calipers as well. It has outstanding toughness and a stiffness making possible excellent folds that hold together during wrapping operations to give packages of neat appearance.

The higher sealing temperature and the wider sealing range of this plastic simplify equipment control. Because of its toughness, the cutoff operation requires heated knives, preferably with accurate temperature controls. The cracking of polypropylene films at low temperature (e.g., freezer conditions) was a negative feature observed in some early applications; but formulations have since been developed to overcome this defect.

Oriented polypropylene (OPP) can be made into very thin films having excellent clarity. When they are coated with a sealant, these materials can be run on wrapping machines without the need for special machine conversions (from polyethylene settings). OPP seals at temperatures lower than are needed for plain cast films but higher than needed for coated cast films.

Coated films can be sealed in printed areas, which are resistant to adhesion if not coated.

Cast films are the least expensive per pound of all the propylene structures, but they cannot be printed without the addition of a heat-seal coating compatible with the ink system being used. Coated oriented film has a glossier appearance and gives a neater, sharper fold, but it is more expensive than plain cast films. The cast, coated, and printed film forms the tightest seals, but is the most expensive of the polypropylene structures discussed here.

Polypropylene has not been much used in thermoformed containers because of processing difficulties. The thermoforming problems of polypropylene, which are in part attributable to thermal degradation of the resin, can be reduced by using stabilizers (essentially antioxidants). There has been developed a method of forming polypropylene just below its softening point; the soft but nonflowing resin is forced into its desired configuration by using mechanical plugs and pressure (Hylton and Cheng 1988).

Polystyrene. This resin is made by polymerization of styrene using a peroxide initiator. It is used in large volume for boxes and thin-walled trays or containers. The raw material styrene, is made from petroleum fractions. Polystyrenes are relatively low in density, have good processability, can form rigid containers, exhibit glossy surfaces, and are fairly cheap. General purpose styrene is rather brittle, cracking readily when distorted, but it is glass-like in its transparency, has high dimensional stability, is odorless, and contributes no off-tastes to the foods it contacts. It has a relatively high water vapor permeability, which is a strong negative for many applications. Modified polystyrene, the so-called "impact" type, exhibits improved strength and is not as brittle as the general purpose variety but its clarity is not as good. Extruded foamed polystyrene is used as the basis for thermoformed cups, plates, etc.

Polyvinyl chloride. Vinyl chloride, the raw material for this plastic, is manufactured from petroleum products and a chloride source. The monomer is polymerized by suspension, emulsion, or mass processes using initiators such as peroxides. The two general classifications are unplasticized (rigid) and plasticized (flexible). The plastic is available in varieties having a wide range of toughnesses and flexibilities. Permeability to water vapor is fairly high for flexible types and and medium for rigid varieties. Vinyl films are recognized as being well-suited to wrapping fresh meat and for shrink- or stretch-wrapping products and containers, including bundle wraps and pallet loads. Films made from polyvinyl chloride are not widely used for bakery products.

For bottles, the glass-like clarity of PVC and its good chemical resistance make it a favorite choice for packaging foods such as cooking oil. PVC sheet is widely used for thermoformed blisters because of its clarity, high stiffness, good impact strength, and favorable cost. As film, PVC has been used for skin packs.

Polyvinylidene chloride. PVDC, or Saran, resin is used in making films and coatings. Chemically, Sarans are members of a family of thermoplastic copolymers of vinylidene chloride and co-monomers such as vinyl chloride, methyl methacrylate, and butyl acrylate. Their barrier properties are dependent on the copolymer content and the amount of additives used to obtain satisfactory processing response. PVDC is a very heat sensitive material that gives off hydrochloric acid when it decomposes. It has outstanding moisture and gas barrier properties, and so is much used in packaging foods such as baked goods, confections, cheese, and meat. For many years, it has been a popular wrapping medium for household applications.

Surlyn. This plastic is one of the class called "ionomers," i.e., resins that are ionically cross-linked. Surlyn is a DuPont trademark for variations of the resin containing sodium or zinc ions, and it will be used as an example of the class. Surlyn resins are derived from ethylene/methacrylate acid copolymers.

The density and processing characteristics of Surlyn are similar in many respects to those qualities of low density polyethylene or ethylene vinyl acetate copolymers. Some of the important qualities of films made from this resin are low temperature heat sealability, outstanding hot tack, formability, toughness, clarity, oil and solvent resistance, and excellent adhesion to other important packaging materials such as nylons and foils. Surlyn can be used in complex structures with polyester (PET) films, paper and foil. These laminations are used for foods, pharmaceuticals, and composite cans.

Additional properties in these film structures include good impact resistance at low temperature, puncture and abrasion resistance, high melt strength, easy deep draw, direct adhesion to foil and paper by extrusion coating and to nylon and other copolymers by coextrusion, low sealing temperature, high seal strength, transparency (with low base), and outstanding hot tack.

Resin Combinations

One of the most valuable approaches for improving the properties of plastic containers is the combining of two or more polymers in a single structure. The combining process can take the form of merging the different substances by physical mixing, chemical reaction, or laminating two or more layers. Often the most desirable characteristics of each component will carry through into the final blend while the undesirable qualities are modified or ameliorated. For example, brittle plastics can be blended with substances that give them greater impact strength, sheets of resins having good appearance and strength can be laminated to materials having low gas transmission rates, etc.

A special nomenclature has been developed by plastic technologists to describe the different combinations of resins. Among the terms in frequent use to describe the combinations are alloys, polyblends, copolymers, multipolymers, coextrudates, and composite structures.

Alloys are composite materials made by blending one type of polymer with other polymers or elastomers. There is, presumably, no chemical reaction between the two or more resins in an alloy. For example, the substance that results when styrene-acrylonitrile copolymer resins are blended with butadiene-acrylonitrile rubbers is an alloy. Also, blends of low density polyethylene with linear LDP, which results in a film of improved toughness (Speed 1982), would be considered alloys.

The term "polyblends" has been used to describe a mechanical (physical) mixture of two or more resins or elastomers. An example would be blends of polystyrene and rubber.

A copolymer is a resin in which two or more differently structured monomers are chemically combined into single chains. The chains may have different molecular weights and, in some cases, even different patterns of monomer repetition, but in general the chemical bonds holding the different monomers together are the same throughout the molecule. A block copolymer is a mixed polymer in which the more-or-less straight chain portion of the molecule consists of regions having repetitions of one type of monomer alternating with regions having repetitions of another monomer — an example might be diagrammed as -yy-x-xxxx-x-y-xx-yyy-x-x- for part of a very long chain, of course, with many variations.

A graft copolymer is a chain of one type of repeating monomer units to which side chains of the same or different monomer units are attached or "grafted." Important examples are the ABS resins resulting when acrylonitrile and styrene monomers are added to polybutadiene latex and graft copolymerized. Another copolymer that has been used in food packages, including some cereal product containers, is ethylene-vinyl-acetate (such as the Elvax resin of DuPont). Features said to characterize films made from this resin include lower temperature heat seals, the ability to seal through product contamination, and high flow in seal areas.

Multipolymers is the term sometimes applied to copolymers having three or more different monomer constituents. There is an almost infinite number of possibilities in this class.

Coextrusion is a process by which two or more different resins are formed into films (or thick layers) simultaneously, and the layers immediately brought into apposition before the final shape is formed. This results in a laminated sheet, film, or container possessing the complementary qualities of two or more components.

Composite structures have become extremely complicated as manufacturers attempt to develop packaging structures that include all of the many properties they wish to achieve, while maintaining the minimum possible cost. For example, the structure of a typical aseptic carton could include seven or more layers, a typical example being (from the inside out): polyethylene/Surlyn/foil/polyethylene/paperboard/ink/polyethylene.

Additives

Most finished plastics contain additives that are not plastic resins but modify some of the properties of the resins making up the basic structure of the container. Among the many types of additives in common use today are colors, plasticizers, antioxidants, flame retardants, and fillers. Some of these additives will be discussed briefly in the following paragraphs to give the package technologist a basis for understanding the specifications he may encounter when procuring packaging materials.

Antioxidants are added to plastic resins primarily to prevent chemical decomposition when the resins are heated (often repeatedly, as scrap is reprocessed) during manufacture of finished film, sheet, or thermoformed containers. Hindered-phenolic antioxidants inhibit degradation by donating their labile hydrogens to the radicals generated during thermal breakdown. Hindered-amine antioxidants also contain labile hydrogens and operate in a manner similar to the phenolic compounds. Phosphite costabilizers perform a similar role, but function by reducing the thermally unstable hydroperoxides to more stable alcohols. Further information on resin antioxidants can be found in the article by Chucta (1988).

Requirements for colorants used in plastics that will contact foods include: (1) Conformity to government laws and regulations; (2) Compatibility; (3) Heat stability; (4) Inertness; (5) Dispersibility; (6) Ease of handling; (7) Plateout; and (8) Resistance to deterioration by light or ultraviolet radiation. Selecting pigments to produce a colored plastic packaging component usually involves consultations between the pigment supplier and the plastic processor. The purchaser of the packaging component may have very little direct input in this selection, since he is usually regarded as being concerned only with the end result.

Filled plastics are resins mixed with a cheap, relatively inert substance (usually in the form of small particles, though sometimes fibers) for the purposes of reducing cost, improving physical properties, or modifying the appearance of the finished container.

Reinforced plastics are materials that contain fibers (e.g., of glass) intended to greatly increase the mechanical strength of the finished object. They are used very little, perhaps not at all, in containers for retail food products, but have some potential for bulk containers, bulk display bins, trays for handling packages of bread or rolls, etc.

Metal Films

For all practical purposes, aluminum foil is the only metal film used by the food industry in laminations intended for making flexible packages. Some attempts have been made to laminate steel films with plastics such as polyethylene and polypropylene, but it is believed that such materials are not commercially available at present.

Aluminum foil is produced by passing strips of the metal between a series of pairs of polished steel rollers, each subsequent pair being placed closer together, until very thin webs are formed. Thicknesses from 0.00025 to 0.0015 inch have been used in bakery product packaging. Foil thicknesses of 0.00030 or 0.00035 inch are commonly used for laminating to plastic films intended for moisture-resistant pouch material. Foils thinner than this are difficult to handle in the laminating process and tend to have an unacceptable number of pinholes.

After the rolling process has been completed, the foil is tempered by a controlled heating process. "Dead soft" foil is commonly used for flexible packages, but trays and the like require a harder foil to give increased stiffness. Unsupported aluminum foil puckers and tears easily, and for this reason, as well as to enable a seal to be made, it must be laminated with other materials when it is to be used as a wrapping film. For bakery product packaging, it is often combined with paper, cellophane, polyester, and polyethylene.

Laminations of aluminum foil with plastics are not much used for bread wrappings except for some of the high priced specialty loaves, but they have found a place in wrappers for other bakery products and may, in time, take over some part of the market now held by other films. A properly designed lamination that includes aluminum foil is virtually impervious to moisture vapor and grease (as always, assuming a perfect seal), and can be very attractive. In some applications, its inherent lack of transparency is a disadvantage, however.

Technological advances have permitted the production of films coated with very thin layers of metal that have either been deposited from the vapor state or by buffing the base film with metallic dusts. These vacuum-coated materials, usually called metallized papers, are being used extensively for candy packaging and for some specialty bakery items. The principal advantage of these films lies in the added attractiveness they contribute to the package. The metal layers do not, however, add nearly as effective a moisture vapor barrier as does aluminum foil, partially because application of a sealing film on the metallized surface can cause "crazing,"or many small irregular cracks in the aluminum layer. This is, apparently, a greater problem if the base film is polyethylene, as compared to polyester. Recent developments (Anon. 1993A) provide ways to at least partially overcome this problem, and some newer forms of metallized film are said to approach the barrier properties of foil.

COMMON FORMS OF CONTAINERS

Among the many types of containers used for foods, bags, pouches, cartons, trays, cans, bottles, and jars probably make up over 90% of the volume. Some of these will be discussed below, but space limitations prevent complete coverage of the group.

Bags

Bags are common forms of containers for retail packages of food. The usual container for bread loaves is a polyethylene bag with a reclosing feature at one end. Plastic bags are also used for many other bakery products as well as for foods not falling into this category. Wholesale packs of buns and other single-serve breads are delivered to fast food outlets and other restaurants in large plastic pouches. Polyethylene bags 1.25 to 2.0 mils thick are commonly specified for sliced loaves sold through retail outlets.

Tin-tie bags, relatively uncommon now, are usually seen as gusseted and flat bottomed bags that have been formed of coated paper or some other relatively stiff film and have their top opening provided with a strip of paper-covered metal. This metal strip serves as a re-closure feature. It has been used, and is still being used to a very minor extent, as a container for cookies.

The two major drawbacks of the tin-tie bag are its high cost and its inability to give good protection to the bag's contents. Its main advantages are the convenient reclosure feature and its old-fashioned image. The usual filling procedure is a dump fill, which leads to a large amount of product breakage if the cookies are tender, but slide-in holders of molded polystyrene (which have been filled with cookies in a separate operation) have been used and solve this problem. A heat-seal below the tin-tie area gives fairly good barrier properties and constitutes only a minor inconvenience to the consumer.

Cartons and Trays

Although bags are economical from the standpoint of material usage and packaging efficiency, they fail to provide the dimensional stability and display areas needed by many products. These functions can be provided by properly designed and produced cartons and trays.

Folding cartons. Folding cartons made of paperboard (which may be coated, laminated, or otherwise modified) are used for many food products. They are especially popular for buns, rolls, cakes, pastries, doughnuts, and cookies. Such containers are nearly always overwrapped with transparent film. They do not offer much protection against moisture vapor loss or uptake, particularly since sealing of large areas of film is often a problem, but they can be satisfactory for fresh products intended to be consumed within a few days after packaging.

For products that need superior protection against moisture vapor transfer but require a paperboard outer carton for graphics, stacking, or some other reason, boxes can be used to enclose hermetically sealed pouches made from moisture-resistant films, giving an improved appearance and considerable protection against crushing. These bag-in-box containers are often made by wrapping the board around the film while the latter is being held around a mandrel having the inside dimensions of a finished box. Other such containers are made by dropping a filled and sealed pouch into a box previously sealed everywhere but the top.

Nearly all paperboard cartons are received at the packaging plant in flat form and are set up (erected) at or near the point of fill. A big advantage of these containers is that they are adaptable to almost any size of operation from small to very large. It is perfectly feasible to buy cartons adapted to hand setup operations, and this is done by many retail outlets such as pizza shops, doughnut shops, in-store bakeries, etc. Larger operations will use automatic equipment for setting up, filling, sealing, and overwrapping.

Folding cartons can be classified as tray style or tube style. Cartons of the tray type are formed from paperboard having the sides folded at right angles and locked, glued, or stitched together at the corners. Some of the varieties of tray cartons are (1) glue-end trays, (2) lock-end trays, (3) roll-over end trays, and (4) metal-end trays.

Glue-end trays have corner flaps that are glued to the inner sidewalls; the flaps may be extensions of one end of the the carton.

Lock-end trays are formed by interlocking tabs that can be inserted into slits in the carton blank.

Tuck-end cartons are made with one extended flap tucked down between two short flaps and the carton body.

Roll-over end trays have extended end panels divided into scored segments; the ends are rolled or folded over.

Metal-end trays are similar to set-up boxes and are formed from blanks having metal supports at the corners.

Two-component boxes, in which a separate top slides over the bottom part, can be formed as two trays.

Each of these varieties has advantages and disadvantages — the amount of board needed to form a carton of a specific volume varies with the type and is one of the major factors determining the cost. Strength of the carton, opening features, and the complexity of the machine needed to set up and fill the carton also vary according to design.

Tube-style cartons are made from a sheet of paperboard that is folded over and then glued to form a rectangular sleeve. The panels forming the ends may be fixed in place by applying adhesive, tucking extensions into folds, or interlocking slit tabs. Folding cartons of the tube type include tuck-end, glue-end, lock-end, and combinations of these.

It is generally recognized that the tray style offers greater strength than the tube boxes. Tray style boxes can also be loaded from the top, an arrangement that is often more convenient than the end-loading requirements of the tube style. Paperboard cartons can be made in many variations of design.

Many of the folding boxes used in the U.S. are finished on a so-called "straight-line folder-gluer," the basic functions of which are to pre-fold the glue lap and the third crease, re-open these two folds, apply glue to the glue-lap, and then fold the second and fourth creases to bring the glued lap into contact with the fourth panel of the box. In its simplest form, the folded box is only a sleeve having four sides and open ends. By adding end panels of appropriate conformation, this sleeve becomes a seal-end, reverse-tuck, or double-tuck carton.

According to Sussenberger (1975), the most economical box blanks to fold and glue have a low ratio of width to length because they can be handled at the highest speeds in the forming machine. Design factors that tend to increase manufacturing costs are: (1) Cartons that require additional and sometimes very difficult folding, (2) Cartons requiring timed folding or timed gluing, and (3) Die-cutting that makes the blank difficult to feed, to fold, or to deliver.

Paperboard trays with or without foil laminates have been used by bakers for many years. This packaging alternative has many similarities in construction and processing to folding cartons. Paperboard trays provide "billboard" sidewalls on which attractive graphics can be printed. Generally, these containers are formed or set-up on the line from stacks of flat blanks.

Ovenable paperboard packages were developed to provide containers in which products could be heated, either in conventional or microwave ovens. Trays made of ovenable paperboard can be of the pressure-formed or flat blank type. Ovenable paperboard does not have the strength or dimensional stability of heavy-gauge foil, and it may turn brown or smoke if the oven temperature is above about 400°F.

Molded plastic trays. Various types of cakes and some other bakery products have been packaged in rigid and semi-rigid plastic trays. The trays may be form-fitting to give support and protection to the cake. Molded trays having several recesses for single cakes or cookies or for rows or stacks of cookies are in wide use. The trays may be flat with raised edges if a group of cookies, croissants, or the like is to be displayed. A heat-sealed film lid may be applied, or a paperboard lid may be held on by a plastic band or by a friction-fit design.

Polystyrene trays have been increasingly used for cookies, partially baked rolls, doughnuts, and cupcakes. Regular oriented polystyrene is as transparent as glass, but the impact resistant version is opaque or translucent as a result of the additives required to give it increased resistance to breaking. Polystyrene trays are thermoformed, that is, heated sheets of the material are drawn by vacuum or pushed by pressurized air into molds and then cut to shape. Most bakeries using these trays buy them from plastic converters since making them in-house requires large investments in equipment. This plastic will not withstand oven heat, so the products must be baked in metal pans and then cooled before they are transferred to the trays. Both impact and regular polystyrene compare favorably in cost with other resins used for rigid and semirigid plastic containers.

Expanded polystyrene has been used mostly for egg cartons, meat produce trays, and beverage cups. It is lightweight and relatively inexpensive. Although expanded polystyrene holds its shape well in the absence of applied force, it does not have much mechanical strength. It is difficult to print on expanded polystyrene, but good results have been obtained with certain kinds of non-impact printers.

Plastic trays and pans for microwave heating and baking are available. Bakery foods currently being packaged in such containers include (among other items) pre-cooked pancakes, waffles, pizzas, sandwiches, sweet rolls, and biscuits, and uncooked brownies, pies, and cakes. Some TV dinners packed in microwavable trays include a portion of cake or muffin batter that bakes as the rest of the components of the tray are being heated. Cake and cookie mixes that have to be stirred with liquid (usually water) are being marketed in combination packages containing plastic pans that can be placed in microwave ovens.

Complex problems are faced when designing a tray that allows the contents to be heated either in a conventional oven or a microwave oven. Coextrusions of two or more plastics and the use of susceptors to control microwave energy distribution are principles that have to be considered. CPET trays are said to be able to withstand temperatures from -40°F to 400°F.

Susceptors are generally composed of thin metallized inserts that heat up rapidly when exposed to microwave energy. If attached to trays in which foods are being irradiated by microwaves, they have the effect of controlling the distribution of heat and intensifying the temperature. Three functions have been attributed to these devices: (1) They deliver resistance heat that causes browning by contact with the food; (2) They serve as multiple reflectors to trap microwave radiation until it is absorbed by the food; and (3) They act as control grids focusing microwave energy where it is most needed. Instead of containing foil, the susceptors being used today consist mostly of polyester film that has been coated with aluminum, stainless steel, or inconel by sputtering or deposition methods (Anon. 1993B).

Foil containers Containers made from aluminum foil, either bent and crimped or drawn into trays, bowls, and the like, are very commonly used for food items. They have the advantage that products can be baked in them, making it unneccessary to transfer fragile baked items (e.g., angel food cakes) from a baking pan to the retail container. Foil containers also offer considerable protection to the product during transportation and handling. The cost of foil trays is often substantially greater than alternative forms of packaging, but savings in processing may offset some part of this expense. The usual closure for foil retail containers is a die-cut piece of printed paperboard secured by crimping the top edge of the foil tray around it. Shrink wrapped film covering the whole pan can also be used, and, in the case of angel food cakes is the preferred method.

Foil pans have been widely used as part of the container for frozen baked and unbaked pies, muffins, sweet rolls, many kinds of cakes (including pound, angel food, and layer), and some other kinds of bakery products. There is hardly any advantage to using foil pans for cookies, except for the brownie type. These pans are particularly suitable for items that must be baked (not just reheated) by the consumer, provided baking is not done in a microwave oven.

The thickness of foil used in these containers ranges upward to 0.0125 inch, or possibly 0.015 inch. For display purposes, and to increase absorption of radiant heat during baking, anodized colorants can be applied to the foil surface. The color can either completely cover the surface or be applied in some sort of pattern, but it is no substitute for printed graphics. The aluminum is made in the hardest temper possible, full hard or intermediate hard depending on the depth of the draw required to form the container. The major strength factor is, however, the thickness of the sheet, not the temper. Design of the pan is also an important strength determinant.

Foil has many properties particularly suitable for pans used in retail distribution of bakery products. The pan can be used for baking the product in conventional ovens, either before or after the product enters the distribution channels. Foil is nontoxic, fairly rigid (at the thicknesses used), well-accepted by the consumer, protects the product well, and gives the impression of cleanliness. Most products can be easily removed from foil pans, although sticking has been a problem with some sweet goods. The water vapor transmission of the foil is zero, but in practice moisture transfer to and from the product is largely a function of the type of lid and overwrap and the way they are applied to the finished container.

According to Antoni (1981), baking and distributing in aluminum foil packaging has the following advantages:

(1) Foil is tasteless and odorless so the flavors and aromas of fresh baked products are not altered.

(2) It is lightweight, providing savings in distribution costs, yet it is rigid enough to protect delicate cakes and pastries.

(3) It withstands wide temperature variations, offering the advantage of a bake-in pan as well as a packaging medium. It can go from the oven to the freezer and back into the oven without altering the integrity of the container.

(4) Foil is an excellent grease, gas, and moisture barrier, repelling all external substances while retaining desired internal components. Because of this characteristic, bake-in aluminum foil pans help extend product freshness; moisture released during the baking process is re-absorbed into the product when it cools.

(5) There are fewer cripples since pan dumping is unnecessary.

(6) No time or money has to be spent on pan washing.

(7) Fewer heavy baking pans are needed, thereby reducing capital expenditures and reglazing costs.

(8) Production delays resulting from the time required for recycling heavy pans are reduced.

Examples of specialty aluminum foil containers were given by Antoni in the article cited above. An improvement in the original foil pan designed for cheesecakes resulted in an ovenable board disk being placed in a circular foil pan with a hole in its bottom; this package is placed on a glass or cup and the pan removed by downward pressure, leaving the board disk to support the cake. Among the many shapes of foil pans designed for holiday use are bells, Christmas trees, hearts, and ovals.

Aluminum foil sheet cake pans permit the manufacturer to decorate and distribute the product in the same pan used for baking, reducing many of the steps required when baking pans made of sheet metal are used. Foil pans with crimpable edges accept plastic domes or board covers – improvements include cake packages having dome covers that can be rotated to open or close the package. Plastic dome covers can be designed to snap over the curl of a pie plate without damaging the crust, allowing the baker to merchandise pies without box overwraps.

One method for packaging large or premium cakes is to use thermoformed plastic blisters combined with board trays or foil pans. The excellent display properties of these packages and the product protection afforded by them often justify the added expense, which can be considerable.

Other advantages that have been claimed for replacing conventional baking pans with aluminum foil containers are: (1) Lower capital investment; (2) Eliminates labor and materials costs of greasing, washing, drying, straightening, and refinishing sheet metal pans; (3) Fewer sanitary problems since pan washing is eliminated; (4) There are no damaged goods resulting from depanning operations; and (5) No labor or equipment is required for depanning.

The negative features of foil are cost, difficulty in getting smooth, neat containers (folds and wrinkles are unavoidable), and non-adaptability to microwave heating.

Specifications for aluminum foil containers will include size, shape, thickness of the foil, whether the side walls are to be smooth or crimped, finish on the exterior (natural, anodized, or printed), and rim construction details. Products are usually, but not always, baked or iced directly in the pan. It would obviously be impossible to bake a cake having more than one layer in the foil pan in which the cake will be distributed, but single layer cakes of the most expensive type are being packaged and sold in the pans in which they have been baked and iced. They are covered with a crimped-on flat paperboard sheet, usually polyethylene coated. This seems to be the preferred package for frozen cakes.

Foil pans for pizzas and pies are sometimes given a black coating on the bottom to make them absorb radiant heat, rather than reflect it. Obviously, the color of the surface adds little or nothing to the absorption of convective or conducted heat, and it is questionable how much radiant heat is created in the floor of a deck oven, for example. Of course, if the pan is suspended (on a rack) in the middle of a household oven, considerable radiant heat strikes the bottom of the pan and a less reflective surface should cause more heating at the bottom of the product when its pan is coated with an infrared-absorbant material.

Perforations in the bottom of pans allow the escape of water vapor as the product becomes hotter, leading to crisper crusts in pies and pizzas. They also reduce the number of concavities in the bottom crust.

Other types of coatings. Foil and metallized plastic make it difficult, if not impossible to detect metal contaminants in filled and sealed packages when the usual type of inspection equipment is used. To a considerable extent, computerized detectors overcome this problem. Other approaches include the use of aluminum oxide and silicon oxide coatings as replacements for metallic aluminum on plastic film, yielding packages that are not magnetically- or electrically-shielding when passed through the fields of conventional types of detectors. The packaging material is also more adaptable to recycling (compared to foil laminates), according to Swintek (1993).

The silicon oxide ("glass") coating mentioned above can be applied to either PET or oriented polypropylene by an evaporation process based on the use of a high energy electron beam. A chemical plasma deposition process is being investigated by at least one company; it is said to be more capital intensive but capable of much lower production costs than the evaporation process.

The coating layer is only a few Angstroms thick, much thinner than foil or most metallized applications.

A manufacturer states that the glass film is slightly amber in color, chemically inert, retortable, microwavable, has been accepted as GRAS by the FDA, and can be printed or laminated. In a Swiss commercial application, the glass-coated film is combined with an outer layer of reverse-printed PET and an inner layer of white pigmented cast polypropylene. This structure is said to have excellent barrier properties against moisture, oxygen, and aroma.

Statements have been made that the glass-coated films undergo cracking when flexed, leaving a layer that is readily penetratable by various vapors. Aluminum oxide coatings are said to be less susceptible to this kind of damage.

Aluminum oxide coatings are applied somewhat differently. In an English method, the plastic film is first metallized in a conventional manner, then "washed" in a patented process that converts the aluminum metal layer to the oxide.

Both types of film are said to be manufactured in considerable amounts in Japan and are being commercially used for various types of foods (e.g, candies, retortable pouches, cups, dairy products, and cylindrical cans). According to a Nabisco technologist (quoted by Swintek 1993), ". . . some of these laminates are getting technically close to metallized [films], but economically they still have a long way to go."

The silicon oxide coatings have a slight yellowish tint, in current versions, but are more transparent than the aluminum oxide films, which exhibit a charcoal gray appearance.

Metal and Composite Cans

This type of container is, of course, one of the major factors in the retail processed food market, and it has been so for over a hundred years. Thousands of varieties of foods have been processed in these packages, with relatively few complaints from consumers and with a remarkable safety record. Although metal cans have never been common containers for bakery foods, a few steamed and baked products have been packed commercially in hermetically sealed tin cans for many years. In regular commercial channels, cans are now mostly used for specialty items such as rum babas, macaroons, dietetic cakes and breads, gourmet cakes, and other limited production items. Containers of this general type have been specified by the Armed Forces for ration breads and cake at least since the second World War. The canned bakery items for rations were originally conceived of as an individual serving (one portion) item, but they have since graduated into a container something like a hermetically sealed steam-table pan.

It has been the custom of some producers to bake fruit cakes in the sheetmetal pan in which they are offered for sale. The can, sometimes with a metal friction-fit lid, is often lithographed or otherwise decorated. Consideration must be given to the effect of oven conditions on the exterior finish, and to the interaction of product and can lining during the baking period.

Molded plastic cans and mandrel-wound paper and plastic cans are in very limited use for baked items and snacks; the usual material for the can ends is tinplated steel, although molded plastic has been used for this purpose. There is a moderately large market for refrigerated doughs packed in the mandrel-wound (spiral-wound, usually) composite tubes with metal ends. These doughs are, of course, removed from their container before baking.

CASES, BASKETS, AND TRAYS FOR DISTRIBUTION

The process of designing cases to be used for containing multiples of consumer packages is considerably simpler than developing the unit packages that will appear on retail shelves. One simplifying factor is that there are much fewer materials suitable for using in cases. The majority of all these containers are made of corrugated board; of course, there are several different kinds of corrugated, but nothing like the hundreds of options for the product package developer. Adaptability to graphics is seldom a problem, and food contamination is a much reduced consideration, since the case never touches the foodstuff in ordinary distribution.

A seven-step approach to corrugated case design and engineering development and troubleshooting was described by Coons (1973):

(1) Determine the goals through critical observation. View all phases from receiving to shipping.

(2) Analyze the container's effect on products, packaging, warehousing, materials handling, inventory, shipping, distribution, marketing, and end use.

(3) Estimate production costs. The basic information is generally determinable from suppliers, plant engineers, etc.

(4) Formulate a comprehensive definition of the problem.

(5) Compare alternatives. At this point, every detail of all possible options must be compared and related to the previously compiled or existing data and information. Possibilities may be found in packaging materials, equipment, materials handling, or warehousing.

(6) Evaluate production improvements that could lead to cost savings.

(7) Establish a precise objective, based on consultations with each of those affected by the new project. With these data, it is possible to design a corrugated container giving the product maximum protection at minimum total packaging cost.

One of the first steps is to check all existing tariff and shipping regulations affecting the product as it will be delivered to the carriers. Sketches, or initial designs, must take into account the dimensions of the number of consumer packages that will be contained in the case – protective internal separators of corrugated or other material may be needed. When prototypes are available, filled and sealed cases will be subjected to laboratory tests simulating all the hazards the container and product will be exposed to. Final designs may be actually be shipped through normal transportation and distribution channels; in these cases the shippers must contain product exactly as it will be delivered to the casing machine when commercial quantities are being packed.

When bagged or boxed bakery products are transferred from the shipping area to trucks and thence to retail outlets, they are usually accumulated in plastic baskets (trays). The contents of one or more trays may be dedicated to a single customer as a means of simplifying delivery and reducing errors. As described by Stritch (1986), there are two basic systems involving these delivery containers. One system uses containers held in racks, while the other system utilizes self-supporting stackable trays.

The design of stackable trays has a major influence on accessibility of their contents during a delivery run and on the total amount of product that can be placed in the truck. It also affects the cost of the tray, its durability, and its versatility so far as the number of different kinds of products it can contain. The efficiency with which a space of given dimensions can accept these trays is apparently one of the most important problems troubling the minds of physical distribution vice presidents of large bakeries. Because of the relatively low density of bakery products, weight is seldom (if ever) the factor limiting the maximum load that can be placed in a truck.

A typical rack and tray system will require 20% or more of the available space for rack shelves, uprights, trays, casters, and clearances, leaving 80% or less for product. The major advantage of these systems is the easy accessiblity of individual items. According to Stritch, a route truck rack three trays wide provides easy access to as many as 80 items in shelf openings at the back and front of the rack, while the stacks of containers in a truck without shelving may allow access to only seven or eight items from the top of the stacks. Better access results when trays with low wall features are used; they allow products to be removed from the side of a tray that has been placed below the top of the stack.

Referring again to Stritch's informative article, we find the following suggestions for improving plastic basket designs: (1) Improve payloads by reducing thicknesses of the sides, ends, and bottoms. Some designs require as much as 25% of the available space for the container itself. (2) Reduce the nested cube to provide more space for back-hauling and container return, and also as a means of increasing space available for the delivery man to maneuver; (3) There should be molded-in pickup and holding points that are quick and easy to locate with automated transfer equipment; (4) Provide truck designs with adequate clearance for faster stacking and unstacking; (5) Design containers that can be washed and dried quickly. Deep pockets and crevices that tend to trap dirt and water should be elimintated. (6) Provide features that facilitate truck loading and unloading; and (7) Eliminate configurations that will snag on other equipment when the stacks are moved into and out of the truck.

ENVIRONMENTAL/ECOLOGICAL CONSIDERATIONS

Up to at least the early 1980s, the package development technologist did not have to give much consideration to the environmental impact or ecological effects of his decisions. This situation has changed dramatically, and it is still changing. Various types of food companies have experienced different intensities of regulatory pressure — carbonated beverage producers have had more trouble than flour millers, for example. These differences are due to the relative visibility of the waste generated by the company, and to selective targeting by public relations experts, paid activists, and other employees of industries that hope to profit from the changes resulting from new legislation as well as to random selection by unaffiliated crazies who feel they have been delegated by some higher being to reform the world.

There is little agreement on how environmentally friendly claims on packages influence consumer buying response. Many market research studies have been made on the subject, and the results from a few of them have been published, but the validity and interpretation of the data remain unclear. Sloan (1993) published one of the best and most recent reviews on this topic. She quoted the following results from five different studies:

(1) 80% of respondents said that protecting the environment is more important than keeping prices down, but only 46% had based a buying decision on environmental factors during the preceding six months.

(2) 85% of panelists considered themselves environmentalists, but only 22% had actually altered their buying habits.

(3) 60% of those surveyed said they were more likely to purchase a product because of its green claims than they were three years ago.

(4) Only 10% said environmental claims never affect purchase decisions, while 16% said they rarely affect their purchases. Presumably, 74% said "sometimes" or "often," though this is not reported in the Sloan paper.

(5) 67% of women and 57% of men panelists said they were more apt to purchase a product because of environmental claims.

It would be worthwhile to have a tracking of the change in attitudes of these panels over a series of years; it might show a decline in reported acceptance of environmental claims as more questions arise about the motivations and rationale of ecologists and their ilk. Most people like to regard themselves, or wish to be regarded by others, as kind, caring, responsible, and unselfish individuals (though few of them are really entitled to be described by any of these adjectives) and will express agreement with policies they believe are compatible with such personality traits. If these policies lose wide credibility, or if the panelists feel they have been duped, they rather quickly change their attitudes. And, it does appear that consumers are responding to information questioning some of the basic environmental concepts (what is a "friendly planet?" – nature is red in tooth and claw, as the poet said).

Brody (1990), writing about the alleged need for expensive and elaborate recycling plans, reached the following conclusions (some paraphrasing has been done):

(1) Solid waste from discarded packaging materials does not represent a threat to human life.

(2) Packaging waste will not generate malnutrition, plague, disease, or any other life-threatening or plant-threatening catastrophe of any living person.

(3) In the perspective of [other] issues in food packaging and the world,

solid waste is not an urgent issue.

(4) Food and cereal packaging waste should be studied and its consequences resolved, but at a thoughtful pace.

(5) Unless interrupted by arbitrary legislation or regulation, plastics will continue to penetrate and substitute for so-called traditional food and cereal packaging because they are more functional, lighter in weight, safer, more efficient, and more economical — and generate less solid waste than their traditional counterparts.

(6) Instant solutions dictated by special interest-driven legislation or regulation can lead ultimately to deeper problems, the fixing of which can lead ultimately to still deeper problems, whose fixing will spiral us even further downward.

It does appear that there may be a trend toward saner regulations as we get broader public participation in environmental debates. Rice (1993) reports that no statewide measures restricting plastic packaging were enacted during 1992, and numerous existing laws affecting plastic packaging were repealed or changed in the direction of moderation.

Regardless of the philosophical, ethical, or logical inconsistencies of legislation and enforcement, the package developer must take into consideration the "greeness" of the container, how it appears in the context of the supposed need for "greater protection of the environment." Top management may assign environmental/ecological investigations and decisions to persons outside the technical group, or, at least, not in the product development group. This may or may not accelerate the development project, depending upon the personality and ability of the assigned person.

It is generally thought that a declaration of "made from X% recycled material" on the package helps the marketability of a product. It is reasonably sure that polled consumers will say they prefer to buy such products. It is not at all clear that this preference will outweigh other choices confronting them when they are standing at the retailer's display — is a recycled package worth one cent or ten cents more in price? Is the wasteful, environmental-damaging, or generally non-green package prettier, funnier, more original, and thus definitely more desirable? Is that fifty-cent coupon on the back of a one-ounce pack of potato chips in a non-recyclable pouch the deciding factor? Package designers generally do not know the answers to these questions and must, in the final analysis, operate to a large extent on gut-feel as to the average consumer's response to label claims of environmental/ecological superiority.

There are important ethical conflicts and potential legal problems in the usage of post-consumer scrap (read; trash/garbage) in food-contacting packages (pre-consumer scrap is not free of these problems, either). One of the questions that arises is, what impurities have been introduced into the new batch of plastic, paper, etc., by the inclusion of recycled materials or by the reaction of both impurities and recycled materials with the new material in a batch? What is the effect on recycled materials of the processing conditions they contact? Koelsch and Labuza (1991) discuss some of the scientific and economic consequences of re-processing plastic.

The FDA issued a publication, "Points to Consider for the Use of Recycled Plastics in Food Packaging: Chemistry Considerations." Other voluntary guidelines have been issued by The Plastics Recycling Task Force. "Recycled food-contact articles must be of a purity suitable for their intended use and must meet all existing specifications for the relevant virgin materials," says Malcolm MacArthur, legal counsel for the Flexible Packaging Association (quoted by Forcinio 1993). Such a statement, made without reference to any quantitative guidelines, and from a person not speaking for the regulatory agencies involved, would be considered by most manufacturers as a risky base on which to construct a policy.

Although some sorting, cleaning, and purification has supposedly been applied to scrap that undergoes recycling, most recyclers do not have sufficient knowledge of the health effects of all the possible contaminants to be able to verify the harmlessness of the recycled product. The potential exists for some very serious problems when these unknown (and, at the present state of scientific knowledge, unknowable) materials come into contact with foods and beverages. Regardless of the purity of their intentions and in spite of any precautions that they take to insure safety, the manufacturer of the re-cycled material and whoever uses it to package foods can be absolutely certain that they will be held responsible (in both civil and criminal suits) for any unforeseen results on the consumer, no matter how many years have elapsed between the production and sale of the package and the ultimate use of the food.

A slightly more favorable situation exists when the food manufacturer generates considerable scrap in the packaging operation and is able to re-use this material in the originating factory. According to Dumbleton (1989) "Most industries have substantial internal recycle rates." This does not conform to my observations made in food factories, however.

The problem of migration from the container to food contents in packages made in part from recycled plastics was described by Begley and Hollifield (1993). Although their article is intended to support the use (or further work on the use) of recycled plastics in food containers, they state, "Even if source controls are used to minimize contamination, some contaminated resin may enter the production stream of recycled resin." Also, ". . . because of the permeable nature of plastic materials and their possible commercial and consumer use as containers for chemicals, these recycled polymers could contain potentially toxic contaminants, such as pesticides and industrial chemicals, which could migrate from the package into the food."

Federal Register notices and news releases reveal that the FDA has up to this time considered allowing post-consumer plastics to be recycled into food packages having only limited contacts with the contained food. There were favorable comments on the use of recycled plastics for foam egg cartons, grocery bags, harvesting crates, small retail baskets for fruits such as strawberries, and bottles for carbonated beverages. It is quite apparent that some of these uses involve direct contact of the food with the recycled plastics and therefore direct contact with any contaminants, toxic or not, that have entered the plastic through whatever route.

The FDA is proposing to establish a process for determining when migration of a substance to food for a food-contact article is so trivial as not to require its regulation as a food additive (does the Delaney clause apply?). It can be expected to take months or years before a resolution of the many problems takes place. The chemistry alone could take several years, and the political resolution decades if not forever.

Designers for food-contacting containers should not decide to use recycled materials unless they have in their hands absolute experimental proof and written guarantees from the manufacturer that the material used in the container is absolutely safe for food use. In my opinion, it is not adequate to say that the reprocessed material is safe because it has been inclosed in a covering or shell of virgin plastic, because the dynamics of migration of toxic materials through these shells or coverings have not been completely elucidated. It seems to be fairly safe to declare the package "recyclable" provided technology is known to exist that could take the used package and process it to make something else, say new packages, road surfaces or underlayment, park benches, etc. A California law defining common terms used in marketing was due to take effect in 1990, but some parts (at least) of it have been held up by court challenges. Recently, one court said the proposed definition of "recyclable" was too vague, among other things, the law said "recyclable" meant conveniently recyclable in the state. As Greenberg (1993) states, "environmentalists have often griped that anything short of closed-loop package-into-package recycling wasn't really recycling at all."

As an example of the problems that may confront firms using the simple claim of "recyclable," we can point to the situation that confronted White Castle recently. This chain of hamburger restaurants had been using paperboard cartons they claimed were recyclable. The Federal Trade Commission said this was "false and misleading" because the amount of food contamination (mainly grease) on the carton makes it unrecyclable in practice even though the paperboard itself can, theoretically, be recycled.

The recycling of aluminum cans has been successful because of subsidies supplied by the aluminum producers in the form of prices for scrap that exceed what the free market would provide and by their support for recycling facilities at every level. The publicity encouraging recycling of aluminum, partly supported by foundations receiving financial contributions from interested parties, is also important, as is the resistance of aluminum cans to deterioration when exposed to the weather.

An important factor enabling the use for food containers of material originating from aluminum cans collected from garbage and the like is the method by which the cans are processed, high temperatures and other conditions encountered by the scrap during smelting effectively driving off virtually all organic contaminants. Similar rigorous conditions may also be present in glass container recycling, but they seldom exist in plastic recycling. A question that may not have been answered with sufficient certainty is, "Do toxic metals such as cadmium accumulate in aluminum scrap through multiple processing steps because of apparently negligible contaminations at each step?"

It is said (Comello 1993) that 41% of soft drink bottles made from polyethylene terephthalate (PET) were recycled in 1992, but most of this reprocessed material goes into non-food uses, as in structural materials for automobiles. Even so, this market for used bottles is not firm, because the infrastructure for delivering PET (and other used plastic feedstocks) is not mature and reliable enough to ensure recyclers of an uninterrupted supply.

BIBLIOGRAPHY

ANON. 1988A. Behind package R&D lies consumer attitudes. Food Eng. *60*, No. 6, 64-65.

ANON. 1988B. Polyolefin containers with flavor, oxygen barrier. Food Eng. *60*, No 6, 58.

ANON. 1988C. Thermoplastic polyesters for dual ovenable tray. Eastman Chemical Products Publ. No. *TR-80*

ANON. 1993A. Cookie caper solved. Converting Mag. *11*, No. 4, 22-24.

ANON. 1993B. Film advances. Baking & Snack *15*, No. 3, 39-40, 42.

ANON. 1994. No green claim for White Castle, says FTC. Packaging World *1*, No. 1, 4.

ANTHONY, S., JR. 1982. Package development. Cereal Foods World *27*, No. 6, 264-266.

ANTONI, P. 1981. Aluminum foil packaging. Proc. Am. Soc. Bakery Engrs. *1981*, 146-155.

BEGLEY, T. H., and HOLLIFIELD, H. C. 1993. Recycled polymers in food packaging: migration considerations. Food Technol. *47*, No. 11, 108-111.

BRODY, A. L. 1990. The current flap about packaging's role in the environment. Cereal Foods World *35*, No. 11, 1127-1129.

CHUCTA, T. M. 1988. An improved stabilizer system for polypropylene. Plastics Eng. *44*, No. 5, 73-75.

COMELLO, V. 1993. Smoother sailing for plastics recycling. R&D Mag. *35*, No. 12, 20-22.

COONS, E. A. 1973. Seven-step approach to corrugated design and engineering problems. *3*, No. 6, 43-45.

COPELAND, M. 1988. Consumers and packaging: A love-hate affair. Prepared Foods *157*, No. 10, 92-93.

DUMBLETON, D. 1989. Solid waste – a case for a balanced approach. Converting Magazine 7, No. 2, 32, 34-35.

FERNANDES, J. G. 1978. Organizing the package development function for OTC product marketing. Package Development & Systems *8*, No. 4, 15-19.

FORCINIO, H. 1993. Environmentally correct '90s spur grocery package recycline. Prepared Foods *162*, No. 3, 87-88.

FRIEDMAN, W. F. 1975. The need for a broadened focus in package design. Package Dev. *5*, No. 3, 18, 20-23.

GATTY, B., and BLAYLOCK, C. 1994. Government decisions to impact packaging in '94. Packaging World *1*, No. 1, 7

GREENBERG, E. 1993. West coasting through enviro-issues. Packaging Dig. *30*, No. 9, 24.

HARADA, M. 1988. A new barrier resin for food packaging. Plastics Eng. *44*, No. 1, 27-29.

HYLTON, D. C., and CHENG, C. Y. 1988. Thermoform polypropylene on conventional equipment. Plastics Eng. *44*, No. 4, 55-57.

KOELSCHE, C. M., and LABUZA, T. 1991. Packaging, waste disposal, and food safety. I. Landfilling, source reduction, and recycling of plastics. Cereal Foods World *36*, No. 1, 44, 46-47, 49-50, 52.

MAKINDE, M. A., GILBERT, S. G., and LACHANCE, P. A. 1976. Nutritional implications of packaging systems. Food Prod. Dev. No. 10, No. 7, 112, 114-119.

MARSILI, R. 1993. Optimizing the value and benefits of packaging films. *3*, No. 8, 63, 66, 69, 70, 72 75-76.

PARIKH, D. R., and KNIGHT, G. W. 1988. Orientation and structure: Keys to PE film strength. Plastics Eng. *44*, No. 2, 47-49.

PIERSON, J. 1993. Form+function. Wall Street J. *92*, No. 92, B-1.

PORNER, F. E., and WHITE, O. L. 1972. Save time and money by running tests in proper sequence. Package Eng. *17*, No. 9, 55-57.

RICE, J. 1993. Environmentally friendly persuasion. Food Processing *54*, No. 9, 29-32.

SCHIER, J. 1974. Developing a package for the short market test. Food Prod. Dev. *8*, No. 8, 76.

SCHOENECKER, L. 1991. Green packaging for food products. Cereal Foods World *36*, No. 5, 454.

SLOAN, A. E. 1993. Consumers, the environment, and the food industry. Food Technology *47*, No. 8, 72, 74-75, 91.
SPEED, C. S. 1982. Formulating blends of LLDPE and LDPE to design better film. Plastics Eng. *38*, No. 7, 39-42.
STRITCH, J. J. 1986. Container designs to reduce distribution costs. Proc. Am. Soc. Bakery Engrs. *1986*, 83-90.
SUSSENBERGER. F., JR. 1975. Reduce folding carton costs by judicious structural design. Package Dev. *5*, No. 6, 25-29.
SWINTEK, B. 1993. Formidable films. Prepared Foods *162*, No. 11, 118-121.
TAOUKIS, P. S., EL MESKINE, and A., LABUZA, T. P. 1988. Moisture transfer and shelf life of packaged foods. *In* Food and Packaging Interactions. Amer. Chem. Soc., Washington, DC
THOMAS, R. H. 1977. Package concept development. Package Development & Systems 7, No.4, 22-25.
VARSANYI, I. 1986. Permeability of polymers in food packaging. *In* Food Packaging and Preservation: Theory and Practice, M. Mathlouthi, Ed. Elsevier Applied Science Publ., NYC

chapter nine

LABELS AND LABELING

INTRODUCTION

As previously stated, the construction and graphics of labels and the method and equipment to be used for attaching them to the container must be recognized as primary responsibilities during the product development project. All too often, this task is left until the last, because it is thought that no firm commitment to label design can be made until the product and container specifications are finalized. This is, in fact, partially true, but preliminary steps can and should be implemented throughout the development process. The Marketing Department will be heavily involved in label design and must be kept informed of any problems that seem to be developing, e.g., if the product dimensions must be changed or if protective container materials are incompatible with some kinds of graphics.

The label may or may not be an integral part of the outer covering of the package. In those cases where labeling graphics cannot be printed on the outer surface of the container, separate labels can be stuck on, wrapped around, or otherwise applied to the outer wall of the container, or, if the outermost layer of the label is sufficiently transparent, inserted on the product or on an interior packaging element in such a position that the consumer can readily see the lettering or illustrative material that makes up the graphics. The physical requirements for labels including the methods for applying them will be discussed in the first part of this chapter, with examples being taken mostly from the baking industry.

There are at least five functions that are expected to be accomplished by the graphics applied to a package. They are: (1) motivating the prospective buyer to choose your package instead of a competitive item; (2) giving the consumer directions as to how to prepare the product for consumption, including instructions for opening the package; (3) informing the retailer and consumer of the preferred (or necessary) storage conditions and expiry date; (4) recording the stock number, place of production and/or packaging, production line identification, and production date, (much of this information being in coded form) allowing the manufacturer to locate specific units in case of complaints or other problems, and for making possible computerized itemization at the checkout counter; and (5) providing the information required by applicable laws and regulations. The first of these functions is the responsibility of the Marketing Department, and so is outside the scope of this book. Setting the storage conditions and drawing up directions for use are technically related functions and should be included in the product development process.

Information mandated by laws and regulations, insofar as it can be separated from technical content, is properly a function of the legal counsel operating in conjunction with various groups within the company including, of course, the product development people. Much of the material required by the government to be exposed on the label, is essentially technical in nature and must originate with (or at least be reviewed by) the scientists and technologists who are developing the product. The regulations affecting nutritional statements, ingredient statements, and other mandatory aspects of label design are described in the next chapter.

FACTORS AFFECTING LABEL DESIGN AND APPLICATION

Some of the factors to be considered in designing a label to meet a particular need are the characteristics of the material of which the label is formed, its physical properties (including dimensions), its esthetic properties, its adaptability to the type of printing expected to be used, and, of course, its cost. Unless new labeling equipment is to be purchased, the designer must assure himself the existing machinery will function properly when supplied with the label. A few of these points of interest will be explored in the following paragraphs.

Labeling Consumer Packages

Many food containers have the necessary graphics printed on the outer wrapper or, if the wrapper is transparent, on an inner component of the package. Some metal cans are formed from lithographed metal sheets and do not require the application of any kind of paper or plastic printed material. Most cartons and virtually all pouch materials have the necessary graphics (except expiry dates and the like) printed on them before they are sent to the packager. All of these graphics and text presentations are, from a legal standpoint, labels, but the latter term is most often used for a separate piece of flexible material applied to the outer surface of the product container. Paper (of some kind) is the material most widely used for labels of food packages, although it is often coated with plastic film or a protective lacquer; other materials include shrinkable polyethylene, polyvinylchloride, transparent polystyrene, metallized film, metallized paper, and laminates of paper with foil and plastic films.

The label must be compatible mechanically and in its dimensions with the container surface and with the device that will apply the label to the container. Necessary characteristics of labels that should be known when writing their specifications include: (1) Dimensional tolerance, i.e., the ranges within which the length, width, and thickness can vary in batches of labels received from the printer – often highly dependent on the method of cutting; (2) Paper weight or caliper; (3) Absorbency, sometimes called the Cobb test; (4) Porosity, which may affect vacuum pickup of paper labels at the label-applying machine; (5) Method of bundling, whether with top or bottom stiffeners, bands, etc.; (6) Shape – are the corners square or rounded, etc.; (7) Dimensions of the area to which the label is to be applied, (8) Compatibility of the label material with the adhesive that is to be applied – the latter often dictated by the type of labeling device that is to be used; (9) Surface finish of the container in the labeling area – affects label "grip" during application and its retention during handling, and (10) Grain direction in paper labels, which is important for certain application methods.

Labels can be stuck to a package by brushing on a liquid adhesive at the time of label application, using either water-based coatings such as dextrin or thermoplastic materials liquefied by heat, or by pressing on a label pre-coated with a permanently tacky adhesive ("pressure sensitive labels").

An important variable of labels attached with any kind of adhesive is the ability of the substrate to achieve a rapid and permanent bond with the container's surface. The initial tack must hold the label firmly enough so that the label is pulled from the position in which it is presented by the magazine of the packaging machine, and it must hold the label on the container firmly enough so that the edges of the label do not spring away from the surface. The initial bond must be retained long enough for the adhesive to set permanently either by cooling, pressing, or liquid absorption.

Adhesives vary in their suitability for different label materials and must, in any case, be tested in a production environment before a final decision as to their suitability can be made.

Not all labels are bonded by adhesives. Sleeve labeling consists of applying a pre-printed plastic film around the container, or a transparent film around a product having some sort of graphics on its surface. The bond is due to the sleeve's physical fit to the container's shape. Non-shrink polyethylene can be used, but shrinkable PVC and stretch polyethylene are now widely used for beverage bottles and the like. These films can be printed either by rotogravure or by flexographic methods after which the web is cut and formed into a tube having a longitudinal heat seal.

Methods for Applying Labels

There is a labeling method for every size and type of food production line, whether it is the home/garage production of cakes, pies, and cookies, or the RJR Nabisco level of manufacture. It will be no surprise to the reader to learn that the capital cost and (generally) the lack of versatility of a labeling plant is related to its production rate in a direct though not a linear pattern. The chief advantages of the very high speeds reached in, for example, labeling lines for beer bottles or soup cans, are primarily lower labor costs and lower floor space requirements.

Manual labeling procedures. A labeling method dating back hundreds of years and still used in very small packaging operations is to apply adhesive to the container with a hand-held brush, and then press the label onto the glued surface. Alternatively, the label itself can be wiped with liquid adhesive. A slightly more efficient method is to coat a sheet of wood, metal, or plastic with glue and press the labels on to it, then pass the sheet to a second operator who removes the labels and applies them to the containers. This method reduces the amount of glue that soils the front edges of the label. It is said that two operators can achieve a labeling rate of 600 to 800 units per hour using these simple methods.

Faster hand production can be achieved by using a machine- or hand-operated roller that revolves in a trough containing liquid adhesive; one operator passes labels through the machine to get a controlled film of glue and then places the label in its approximate position on the container. The label is correctly positioned and pressed into place by a second operator. Estimated speeds for a team of one glue applier and four affixers is 40 to 50 containers per minute.

Pressure-sensitive labels supplied on sheets or rolls are suitable for hand labeling operations. One or more conveyors for carrying the packages to and from the worker will be necessary, as the minimum equipment needed for any acceptable level of efficiency. For cylindrical bottles or jars having smooth surfaces, it may not be necessary to take the package off the line, the operator applying the label as the jar is passing his station. Other shapes may require some sort of positioning equipment. Large pressure-sensitive labels or labels cut in fancy designs may lead to the generating of much scrap as a result of wrinkling, folding, and other malfunctions at the application point.

Obviously, these hand methods are totally inappropriate for any kind of large volume production. They might be suitable for short runs of experimental products required for use in sales or marketing tests, or in special emergency situations. The semi-automatic methods that are available can give reasonably good production rates and are particularly suitable for factories processing many different types and shapes of containers in relatively short runs.

In the simplest type of semi-automatic glued label appliers, a machine picks a label off a stack, smears adhesive on the back of it, and then applies the label to the container. The containers are manually placed in position to receive the label, and then manually removed. The initial tack of the adhesive must be greater for this operation than for hand application. Maximum speeds of 60 containers per minute are said to be achievable under ideal conditions.

Simple semi-automatic equipment works well with most pressure-adhesive labels. In its most elementary form, the machine will pick a preprinted and pre-cut label from a large roll and roll it on (or blow it on) a container that is being conveyed past the applicator. The containers may be placed on, and taken from, the conveyor manually or automatically, depending on the degree of sophistication of the device.

Automatic labeling. Nearly all high speed automatic labeling systems are based on four manipulations or operations: (1) Labels are fed from a magazine or from a roll; (2) The label is picked up by suction cups, in most cases, although blowing on by compressed air or pick-up by an adhesive surface can be used; (3) Adhesive is applied by rollers either to the back of the label or to the container, either in complete coverage or in a pattern that only partly covers the surface (as in strips, dots, or some other pattern); and (4) The label is pressed onto the container by belts, pressure pads, brushes, or compressed air. Thermoplastic labels and pressure sensitive labels require slightly different techniques.

Conveyors are important and integral elements in a packaging line; their efficient functioning is critical to the performance of the labeling operation. The basic concept of conveying to and from the in-line labeler is that containers must not undergo any abrupt changes in direction during their movement. Containers such as bottles or boxes arrive at the labeler contacting each other and are separated there by a feedscrew that delivers them to the labeling area at the necessary rate and spacing. The labels may be applied either at the feedscrew or immediately after the container leaves the feedscrew.

Application methods used with magazine-held or roll-fed labels include vacuum or mechanical transfers with cold glue or hot melt adhesives. Afterwards, the containers, if flat, will pass between dual belts that press the label against the package to encourage adhesion. If the shape of the container is irregular, individual compression pads that conform to the container and label shapes can be used. Round containers rotate through a label roll-on station consisting of a rotating belt and an opposing sponge pad. See Schiessel (1984) for more details.

Many modern labeling machines continuously monitor container flow by a series of sensors so that line-speed fluctuations can be detected quickly and warning given of broken, misplaced, or missing containers. So-called zero-pressure conveyor systems are available. They combine the containers into a single line before they enter the labeler and accumulate them after they leave the labeler. They can be expected to reduce breakage or crushing and to permit easy insertion or removal of containers at any point, as for inspection and quality control sampling.

Changeable information, such as expiration dates and batch numbers, is routinely applied on-line by embossers, perforators, lasers, or ink imprinters. Imprinters can be combined with label applicators in a single station.

Label Substrates and Adhesives

The label itself is basically defined by the substrate, the printing inks and printing process, and other materials and treatments applied by the converter (label manufacturer). The substrate is the material on which all the other materials are applied, as for example the paper that the converter receives in unprinted roll form. The adhesive must also be considered an integral part of the label, whether it is applied at the point of application or by the converter who manufactures the label stock, because of the interactions of the label and adhesive both at the time of labeling and during the shelf-life of the product.

Substrates. Many label substrates are available, the options including paper, plastic, foil laminates, and combinations ranging from the heaviest embossed stock down to printed transparent polyester film of 1-mil thickness. Specially coated labels can be produced to protect the graphics from moisture, abrasion, grease, or chemicals. Labels can be die-cut in almost any shape, and they can be applied to almost any part of the container surface, including recessed areas.

Paper is both an economical and a versatile material. It can be printed in many colors, embossed, cut, and otherwise "converted" to give attractive and readily machinable labels. It also affords some (a relatively small degree) of protection against breakage of glass containers. Because it is ordinarily susceptible to scuffing, moisture penetration, and other environmental damage, paper is often overprinted with a lacquer or laminated with a transparent plastic film to protect the decorative material that has been applied to it.

Many plastics are found in label substrates, sometimes as part of a complex lamination, while at other times serving as the basic label material. Even expanded polystyrene has found a place as a label material; it furnishes substantial protective and safety values when placed on carbonated beverage bottles.

Adhesives. The success of any labeling operation involving the application of liquid (including melted) glues is strongly dependent on the properties of the substance used to affix the label to the container. There are two major categories of these adhesives:

(1) Hot-melt glues, normally procured as pellets of various sizes, shapes, and colors – they must be liquefied by heating them in the labeling equipment.

(2) Adhesives to be applied at room temperature. They are delivered as pastes or liquids in drums, or as bagged powders to be mixed with water.

Several kinds of permanent and removable adhesives are stocked by suppliers of these materials. The following discussion applies to the use of adhesives for case sealing, paper laminating, and certain other functions, as well as for labeling.

Cold setting adhesives of the so-called natural type are mainly based on water dispersions of either modified starches or animal materials (casein from milk, etc.). The commercial type of starchy adhesive may be produced from raw flours and starches, from treated or modified starches, or from dextrins. In nearly all cases, they contain additives to give the product desired qualities such as quick tack and flexibility. These glues are regarded as easy to use and simple to clean up; they generally have good adherence to paper and good heat resistance. They are also relatively inexpensive and are used extensively for labeling many kinds of packages. Casein glue is produced by acidifying the protein component of milk; it has good tack and adhesion, and is resistant to softening in cold water after it dries.

Synthetic cold adhesives are typically emulsions of plastic resins. They are often based on water suspensions of polyvinyl acetate, to which are added protective colloids such as polyvinyl alcohol or hydroxyethylellulose. Various other additives (plasticizers, fillers, solvents, etc,) may be used for modifying the properties of the basic adhesive substance so that it is better fitted for its intended purpose. These glues can be formulated to be low in color, taste, odor, and toxicity. Case sealing is the largest single use for such adhesives, but a substantial quantity is also used for labeling (Kaye 1984).

Hot-melt adhesives do not use a volatile solvent such as water, but rely on liquefaction by heat and subsequent solidification by cooling to perform their function. Because heat can dissipate faster than volatile liquids, bond formation is rapid, resulting in shorter set times. This property is particularly important for high-speed operations in which rapid tack development is essential. Most hot-melt adhesives for labeling use are based on ethylene vinyl acetate (EVA) copolymer combined with a low molecular weight resin and wax. Each of these ingredients provides certain necessary properties to the final adhesive product.

Other hot-melt systems are based on branched low-molecular weight polyethylene and low molecular weight resins. Polyethylene systems generally have less hot tack, shorter elongations, and weaker peel values at elevated temperatures but they are usually lighter in color, machine better, have greater shear strengths at elevated temperatures, and are more resistant to grease and oil, as compared to the EVA copolymers. Additional types of systems are under development.

A disadvantage of hot-melt adhesives is their limited heat resistance. Thermoplastic materials will lose a considerable amount of cohesiveness at only slightly elevated temperatures (Quinn 1984).

The use of deadened adhesives, folded stock, or patterned gum allows part of a label to be peeled off and used as an instantly redeemable coupon for in-store promotions (Riley 1984).

Machines for Applying Labels

The type of label to be used and characteristics of the container to which it will be applied define, to a large extent, the specifications of the equipment that will affix the label to the package. A few of the options will be described below.

Applicators for heat-seal labels. Machines for applying heat-seal labels will include, for example, the following steps: cutting the label from roll-stock with heater "shoes," holding the label on the heater shoes by vacuum until it has been moved to the point of application, releasing the vacuum, and applying the label to the package. While the label is in contact with the heated shoe, the adhesive on the label becomes activated. The adhesive side of the label is separated by the paper structure from the heater shoe so the latter will not accumulate adhesive. Heating elements never come into contact with the package, so that "burn- through" does not occur. Imprinters for applying variable copy, such as date codes, to the labels can be included as an intermediate part of the station.

Applicators for glued labels. Automatic machines for affixing glued labels are of two kinds, rotary and in-line. Containers entering a rotary labeler are separated and spaced by a feed screw, which delivers them at the required rate and separation to the infeed star wheel. As they pass through the infeed star conveyor, they are detected by a proximity sensor that activates the label transfer process. In one design, the label magazine moves forward, allowing the applicators (which are coated with a thin film of cold glue) to contact the back of the label that is on top of the stack. The label with its backing of glue is transferred by the gripper cylinder to the container, which has been moved into position. Immediately after the label is applied, the container makes a series of turns to expose both the leading and trailing edges of the label to the wiping device. If a back label is required, the container is turned to face the back gripper for its application. Rotary labelers are available in different flow configurations to meet requirements of different line architectures.

Applicators for pressure-sensitive labels. Self-adhesive or pressure-sensitive labeling is a practical option for food packaging in many cases, though the materials costs are usually substantially higher than for glued labels. Self-adhesive labels are delivered from the printer with adhesive already applied and the sticky surface protected by a silicone-coated web. When peeled from their backing, the labels are ready for immediate application to the package. Consumer versions of such labels are familiar to all of us.

Commercial labels are generally produced in roll form to simplify loading into labeling machines, but they can be obtained with a fan-folded backing web, an arrangement that facilitates attaching of the following end of a strip to the next stack to make continuous feeding possible. Modern label adhesives of the pressure-sensitive type have reasonably good shelf-life, but overlong storage will in all cases eventually lead to deterioration of adhesive properties to such an extent that the labeling machine will not function properly.

As might be expected, automatic equipment for applying pressure-sensitive labels is somewhat simpler than equipment that must store and disense adhesives. Certain designs of continuous rotary motion devices permit application at speeds in excess of 1,000 label inches per minute while maintaining registration accuracies of about 0.03 inch.

In general, functions of machines for applying pressure-sensitive labels are reduced to properly positioning the label and container, removing the label from its backing, and attaching the label to the container. Equipment for applying these labels operates on one the following four principles: (1) Blow-on machines that peel the label off its backing and hold it in place on a vacuum grid until a container passes in front of the grid, at which time a blast of air transfers a label to the container surface; (2) Wipe-on devices that cause the container to stick to a loose edge of the label and peel it off its backing — the rest of the label is then wiped on or rolled on to the container in a continuous motion; (3) Tamp-on equipment that peels the label off its backing and the mechanically pushes it onto the container surface, a technique particularly suitable for containers having recessed panels, or in other applications where the label must move a relatively long distance (such as an inch) before it contacts the container; and (4) Tamp/blow apparatus that combines two types of movement — mechanical thrusting and air blast — for special situations where greater distances are involved or other difficulties exist.

IMPRINTING LABELS WITH VARIABLE INFORMATION

Although the food packager will probably want to obtain labels, packages, and cases pre-printed with the copy that is expected to be uniform for all the intended uses of the labels or containers, there is still a need for printing variable information on containers going through the packaging line. Types of imprints that must be varied too frequently to be placed on pre-printed labels include batch codes, production and/or expiry dates, and production line (or shift) identification.

Pre-printed labels give many more options for graphics than do labels that are printed entirely on the line, even though on-line imprinting equipment has been remarkably improved over the last two or three decades. Multi-colored, well-registered imprints with half-tones or other attractive illustrations can easily be produced on the equipment of label manufacturers, at a quality level it is quite impossible to achieve on the line or even in the print shop of most food companies. The food manufacturer, however, will often find it inconvenient and costly to stock large quantities of dozens or even hundreds of different labels and so must rely to some extent on in-house imprinting.

For producers of private label goods, this problem is compounded by the number of different distributors being served, each of whom must be accommodated with special labels. If a number of different private label brands are being produced, the printer may be able to gang print several different labels from one plate, providing substantial economies in plate preparation and printing. Matching quantity produced with the largely unpredictable quantity to be used is the main problem in gang printing.

Manufacturers often find it is both convenient and economical to buy large quantities of generic labels, usually containing the company's name, logo, address, and other permanent information, but omitting (for example) the product name, net weight, and ingredient statement. Small bakeries (and some larger ones, too) find this approach essential in order to meet labeling laws and customer requirements. On these pre-printed labels will be imprinted the name of the product being run at the time the label is applied, the product's ingredient statement, net weight, and (sometimes) its price.

It is important for marketing purposes that any information printed on the packaging line present to the consumer a legible and reasonably attractive appearance. A smeary, indistinct, and poorly oriented message appearing to have been imprinted by a rubber stamp wielded by a careless worker makes the prospective customer uneasy about the materials and workmanship that went into the foodstuff itself. Fortunately, there are machines that can give neat and legible imprints, either in the plant's print shop (if there is one) or on the packaging line.

Nearly all food manufacturers will want to imprint on the package the date of manufacture and/or an date of expiration. Often, it is also desired to imprint a bar code identifying the manufacturer, product, and production line. This is obviously the type of information that must be printed on-line, although it is not always necessary to print all of it on the label — production dates and batch numbers can be printed directly on the lids of cans, for example.

Canners have for decades embossed on can ends the manufacturing date and other information such as plant numbers. This information was almost always in a form unreadable by the consumer, and often, because of inadequate embossing techniques, not readable by anyone. Metal deformation caused by the punches had a weakening effect on the can structure and occasionally exposed spots of bare metal to contact with corrosive foodstuffs as a result of destruction of the liner lacquer. This was seldom observed, however, if the embosser was properly adjusted. Embossing of identification information on the metal lids of glass jars (as for pickles, etc.) was also widely used. At the current stage of technology, laser coding appears to be the method of choice for metal cans and can lids.

For other types of coding, as on fiberboard boxes, cartons, and trays, and on some plastic pouches, coding can be accomplished by rubber roller types of contact ink imprinters, or by non-contact ink-jet printers and laser marking devices. Depending on the circumstances, any of these methods can give satisfactory results.

The inked roller machines usually have the smallest initial cost, but the amount of ink used is probably greater than with ink jet printers. Although the codes imprinted by both ink jet devices and laser machines can be changed with extreme speed if they are controlled by computers, the simple changing of a few letters or numbers on an ink roller also requires very little time, so this is may not be an important factor in many operations. If the message requires more than a few symbols, however, it is probable that the roller type of device would not be very practical.

Imprinters for Packaging Lines

In recent years, there has been a steady improvement in electronic printing methods for pressure-sensitive and rollstock labels. These machines are controlled by computers having permanent and/or reprogrammable memories allowing retention and utilization of many pieces of information that must appear on the label, as well as instructions for guiding machine actions. Some types of equipment will print labels and then transfer them onto products, while others deliver the printed labels in roll form suitable for placing on the applicators. More advanced electronic printing devices can be combined with preprinting on a letterpress unit, which gives certain esthetic properties difficult to achieve with ink jet, rubber roller, or laser marking.

According to Fraser (1984), manufacturers considering procurement of an imprinter for their packaging line should specify: (1) The size, minimum and maximum cut-off lengths; (2) Whether the marking will be done on the line or in a separate operation; (3) Whether the package will contain the product when the marking is done; (4) If a registered mark is required; (5) The legend and the number of characters it contains; (6) The type of materials handling system that will be used to move the package; (7) Installation requirements (space limitations, environmental problems, etc.); (8) The range of speeds of the packaging line; and (10) the size, permanence, graphic quality, and legibility desired in the printing

The three major types of devices used on a packaging line for placing codes, logos, and other legends and decorative symbols on labels and containers are contact printers (several kinds), ink jet systems, and laser coders. Contact coders are the oldest and probably the least expensive of the three types. Contact printing methods include wet ink printing, hot stamping, and instant dry ink systems. Direct wet printing uses the familiar transfer of pigmented paste or fluid to the receiving surface through the agency of a raised printing design — rubber stamps are close relatives.

Offset wet printing introduces an intermediate printing blanket of cylindrical shape to transfer the ink from the raised design to the package surface. The offset method frequently gives better definition. Disadvantages of wet ink systems include the need for a drying time, possibility of smudging, and general messiness.

Hot stamping, or foil printing, uses dry inks bonded to a thin web of carrier film. When the film is passed between heated metal type and the container surface, ink is transferred to the package. This method gives a clear, sharply defined imprint of dense color, but it is generally more costly (both for equipment and supplies) than the wet methods.

Dry ink systems are fairly recent developments. They use ink that is heat-transferred from an impregnated roller. The cylinder is dry to the touch and will not splatter or spray ink at high operating speeds and avoids smearing and related problems.

Imprinting by ink jet. Ink jet systems are used for time and date coding, consecutive numbering, logo design printing, shift coding, lot and weight information, and bar codes. They function by projecting very small electrically charged droplets of ink from a pressurized reservoir onto the surface that is to be printed. The path followed by the droplets is controlled by high voltage plates that deflect the droplets according to a programmed pattern.

The main advantages of jet printers over contact printers are that the former can print on nearly any shape or surface, including fragile or unsupported surfaces, and are compatible with computer-controlled printing (Lingle 1988). Ink jet systems can be classified as either small character systems for on-product coding or large character printing devices for case coding applications. Either type offers extensive message storage capabilities and can operate at speeds up to 2,000 characters per second.

Inks to be used in jet printers must be formulated to meet the special requirements of this equipment, nontheless they are available in many different variations. Jet inks can be manufactured in many colors for printing on porous and non-porous surfaces. For some non-porous surfaces, thermoplastic fluids that are fast-drying and abrasive-resistant can be applied — these are suitable for metals, glass, film, polystyrene and some laminates.

Imprinting by laser. Laser coders are inkless alternatives to jet systems. The lasers in these machines emit intense and narrowly focused beams of light of nearly homogeneous wavelength, usually in the infrared range. When the beams strike a surface that absorbs them, an almost instantaneous heating takes place in a very small area. These devices can create a permanent mark on glass, paper, and many kinds of plastics and plastic-coated materials. The light energy either removes a thin layer of material such as the ink on a paper label, etches the surface of materials such as glass, or causes a color change in plastics and the like. The depth of penetration is small, under normal conditions of operation, so that all of the energy in a millisecond impulse is concentrated in a microscopic volume of material. Thus, the usage of energy per unit is small, but it is nonetheless a cost that must be considered.

A big advantage of laser codes, properly applied, is that the "printing" cannot be erased, rubbed off, smudged, or washed off. It can, of course, be obscured by dirt or other unwanted materials that accumulate on the package surface. This permanency is of considerable value in retaining product identification during difficult or damaging storage and transportation conditions.

When a laser coder is in place on a packaging line, its light beam is triggered by a sensor that detects the presence of a package in the right position. The light pulse is directed on to the package surface by a mask or stencil and a lens so that it forms the letter or symbol required.

Lasers are normally used only for on-product coding, not for label imprinting, which can be done better and faster by other methods. Current versions of this equipment are not as versatile as ink jet coders but may have other advantages.

According to one authority (R. Macklin as quoted in Anon. 1993), new dot matrix laser coders do not require masks for imaging code information; they can adapt to very rapid programming and code changes from a host computer, or from a keyboard attached to the coder. They can form characters as large as 0.6 inch and in one-, two-, or three-line formats. Macklin continues, "Dot matrix laser coders employ sealed carbon dioxide lasers, so no external gas supply is required. The latest systems also have eliminated the need for external cooling water by adding an integral water-to-air heat exchanger within the coder." Such systems are powered by 115-volt lines; they do require a supply of compressed air to prevent dust from fouling the lens.

Polypropylene and polyethylene are substantially unaffected by laser energy since they do not normally absorb enough of it to be noticeably changed, and, if the energy is absorbed (because of colorants in the plastic, etc.) the resin tends to melt rather than discolor. Recent developments in pigmentation of PE and PP are claimed to improve the laser imprinting response of packaging materials made from them. Bare metal surfaces are also generally unsuited for laser marking since they tend to reflect rather than absorb the radiant energy. However, coated or anodized metal surfaces are said to be satisfactory substrates under certain conditions.

Ink jet printers are said to require more maintenance than laser coders, and they certainly require more materials (ink) than lasers (which require none). On the other hand, ink jet printers can print large dark figures and letters that are easier to read and less subject to misreading than are the symbols imprinted by lasers. The relative position in the capital costs range of these two types is not entirely clear. Environmentally, the position is not entirely clear either — the ink jet devices emit solvent fumes while the laser devices usually generate noticeable amounts of smoke and other fumes from burned packaging material.

Variable Information that Should Appear on Labels

Among the many types of information a food packager may find it necessary to imprint upon labels are code dating and the universal product code (UPC). The latter symbol will be usually be applied at the time the label converter runs the substrate through his presses, but there are some instances when it must be imprinted on the packaging line. Discussions of the methods for determining or establishing the code dates and UPC are given in this section.

The UPC code, which is only one type of innumerable possible types of machine readable identification symbols, will be described in considerable detail later in this section. Other forms that may be encountered by label designers are the "interleaved 2-of-5 (I 2 of 5)," "code 39 (3 of 9)," "code 128," and "codabar."

The interleaved 2-of-5 system uses only numeric characters, the code is of variable length and must have an even number of characters, bi-directional scanning is possible, and the quiet zone is 10 times the width of the narrow bar. In this bar code, an even number of characters are required, and the characters are interleaved using the bars to represent data characters occupying the odd positions and spaces for the even positioning. Each data character consists of five elements, two are wide and three are narrow.

Code 39 symbols can be of variable length and they incorporate alphanumeric characters (a significant advantage in some applications). They permit bi-directional scanning. This type of coding is seldom used on food packages.

Code 128 is so named because it has 128 characters. It is more compact than code 39 symbols, and it can also be scanned bi-directionally. There is no check digit because it is a discrete, self-checking code. Each character is represented by a stand-alone group of four bars with the three spaces they enclose.

Code dating . For many years, food manufacturers have been applying date codes to consumer packages. The manufacturing date is almost always present, and, in addition, an expiration date is often imprinted to give the time at which it is expected the product will not be in satisfactory condition to be consumed. Expiry dates easily read and understood by the consumer are valuable purchasing incentives — consumers understandably want to know whether the food they are going to purchase is likely to be stale or spoiled. For this reason, many manufacturers have begun the imprinting of either the manufacturing date or (more often) an expiration date on the product at the time it goes through the packaging line.

Calculating the "Use by" date to be imprinted has to be based on the addition of a fixed number of days to the the manufacturing date. Normally, this factor (fixed number of days) will not vary because of the time of year, but for sensitive products such as chocolate-coated marshmallow cookies or the like, seasonal variations in temperature (and humidity) may have to be taken into account.

The food will not necessarily be unhealthful or spoiled at the expiration date, but its sensory and functional properties will usually be so much below par that the average consumer will be dissatisfied with the product's quality. In all cases, expiry dates assume storage and handling conditions will be within recommended ranges. It is obvious that ice cream stored at the preferred temperature of 0°F, for example, will remain acceptable for hundreds of hours longer than one stored at 100°F. A chocolate bar with a normal shelf-life of several months at 70°F will become unacceptable to the average consumer within an hour if stored at 100°F. Salt or sugar will remain in satisfactory condition practically forever, regardless of temperature, if protected from moisture absorption, but will become unusable (because of lumping) within a short time if exposed to high humidity. Pasteurized milk in an unopened consumer container of plastic-coated fiberboard will be drinkable after ten days (or more) if held continuously in the 35°F to 40°F range, but, once the container is opened the milk has a substantially shorter life because of spoilage by microorganism contamination from the atmosphere.

The preceding examples point out the large contribution made by subjective and unknowable factors in determining the shelf-life that is to be used to in calculating expiry dates. No universal guidelines can be given in the present discussion, but it is reasonably clear that a knowledge of existing storage and handling practices, as determined by reviewing actual examples, must be considered as one of the most important factors to be used in calculating "Best to use by" or "Use before" dates intended to be read by the purchaser.

Universal product codes . The UPC is a 12-digit code, the first digit being a "number system character" (NSC) designating the product type, the five following digits being a "manufacturer identification number (MIN)," then a five-digit "item number," (IN) and finally a single digit "modulo check character" based on a calculation applied to the other numbers and serving as a means of checking the accuracy of the reading device.

The number system character/digit appears on the far left of the eye readables; it designates the type of product. Numbers 0, 6, and 7 all indicate "Food and General Merchandise." Number 2 indicates "Random Weight Items," number 3 "Drugs and Pharmaceuticals," number 4 "In-store Labeling," and number 5 "Coupons," while the remainder are "Reserved."

The manufacturer identification number, being the five digits following the number system character, is assigned to the user by the Uniform Code Council. Different manufacturers can be assigned the same MIN, being distinguished from each other by different NSCs.

The IN identifies the product. The manufacturer assigns each uniquely specifiable product a separate item number.

The number code can appear as the common form of numerals, presented in whatever font it is convenient to use, or in other forms. We are most familiar with it as the bar code appearing on many labels (as well as elsewhere). A bar code is simply a machine-sensable, computer-translatable arrangement of, e.g., vertical bars (or straight lines), the position and thickness of which contain the information of interest – the heights of the bars are irrelevant as long as they are continuous through the strip read by the sensor. In the usual presentation, a set of bars and spaces (two of each per digit) will have underneath them the number they represent. Each set actually consists of seven "minimum data element widths" allowing the bars (and spaces) to convey information in an efficient form.

Among the essential non-numerical elements are a "left guard bar pattern" and a "right guard bar pattern" that orient the sensing device as to the direction in which the pattern is to be read. In addition, there is a centering pattern dividing the numerical information into right- and left-hand segments. Outside the guard patterns is a blank zone to isolate the pattern from the clutter or static of the rest of the printed material on the label. The bar pattern may be of almost any size, so long as it is conformable to the inherent limitations of the scanning device; it is the relative thicknesses not the absolute thicknesses of the bars that conveys the information.

Slot scanners, such as are found at retail checkout counters, are designed so that at least one beam will pass through each half of the symbol regardless of the direction in which the UPC pattern passes over the sensor. This allows the code to be read in an omnidirectional manner. Other types of scanners are the wire connected wand/pen designs, and the wireless cassette modules.

Although present-day scanners do a remarkable job in reading bar codes having various printing defects and presented to the sensing unit in unusual positions, there are some precautions it is essential to observe when developing graphics containing these information packets. Among these is the size of the symbol. The larger the bar code, the less critical is the orientation of the package as it is moved across a store scanner. As the symbol becomes smaller, it is more difficult to retain the bar thickness proportions using common methods of printing (such as flexography). The recommended "nominal size" requires an area of 1.469 inches wide by 1.02 inches high. The UPC size is called "the magnification" and can range from 80% to 200%. If the available space is smaller than 80%, various strategems have been suggested to cope with the situation, truncation (reducing the height but not the width) is one approach – it leads to scanning problems.

Black is the much-preferred color; reddish hues should be sedulously avoided. Metallized or foil backgrounds should be overprinted with white in the area where the bar code is to appear.

LABELS WITH SPECIAL FUNCTIONS

The susceptor plaques sometimes placed on or in baking containers intended to be heated in microwave ovens can be obtained as rolls of pressure-sensitive labels and applied to the pan by automatic labeling machines. Die-cut shapes suitable for any specific requirements can be supplied.

Time/temperature labels that visually or electronically indicate remaining product shelf-life have been developed (Morris 1988). In one form, an adhesive label packet contains two liquids separated by an internal seal. One of these cells contains a green-colored lipase and pH indicator solution, while the other contains a fat absorbed on a solid carrier. The properties and location of the reactants are selected so as to give a constant reaction rate. To activate this monitor, the internal seal is ruptured by manual or automated pressure. The reaction then begins and continues at a rate determined by the temperature. As acidic products of the reaction accumulate, the color of the indicator changes, e.g, to brown, to indicate that time and temperature conditions have been sufficient to cause unacceptable deterioration of the product packaged in contact with the indicator device.

LABELING SHIPPING CONTAINERS

The cardboard boxes and shipping trays used to contain several packages during their transport from the factory to wholesale or retail customers (or outlying stores) are usually labeled to indicate product and destination. Other information is often included on the labels. If the distribution system to be used for a new product involves these carriers, their limitations must be considered when setting up the project specifications.

The following discussion of a typical basket labeling procedure in a wholesale bakery is closely based on an article by Richter (1984). When customer orders are received, they are used to generate a production order, route orders, customer order summaries, and label lists. The label list tells how many shipping labels need to be printed for each route. The day before the load goes out, customer load summaries are checked with production to assure that all needed items will be available in time. Load label lists are used to assign the trailers, to determine if overloads exist, and to furnish the basis for printing pressure-sensitive shipping labels to be attached to the trays. The labels, which are prepared in many cases by a separate automatic printer but could also be made by computer, include a five-digit code identifying the customer and route. Labels for each trailer load are wound on a separate roll.

Filling of orders is done by a team of two persons. One man feeds the cake and cookie items for a route onto the shipping conveyor while the other transfers the sweet rolls and pies. The route order is placed in the last tray of the order as a marker. Roller conveyors in the storage area are assigned to various products in the same sequence as they appear on the route orders; this assists the pickers to work quickly and accurately. Filled trays pass a labeling station where a label is manually applied to each one. The last label for a route will go on to the tray containing the route order. If the number of trays and labels is not the same, a check of the order is required.

After the trays are labeled, they are transported to the truck, which is manually loaded. In the operation being described, one man stacks the trays four high and turns them so that all of their labels face the back of the truck. Two more workers place the stacks of four in the truck. Since the capacity of each trailer is known, the total number of trays in a load can be verified and compared with the orders as a final check. Each day, a summary report is made containing data on which trucks were loaded, how long it took to load each truck, whether the total number of trays on each truck was correct, and whether the number of shipping labels was identical with the number of labels.

When the drivers make their deliveries, they check the total number of trays left at each location with the route order for that location. Overages, shortages, and damaged items are reported. The number of empty trays picked up is recorded. The drivers' reports and the daily shipping records are summarized in a weekly shipping report. Errors reported by shipping and drivers are compared. Damaged merchandise, loading time, number of trays loaded, and the number of trays loaded per hour are also included in the weekly report.

Drivers' reports are used to generate a tray report each week. This shows how many trays each customer received and returned every week for the previous nine weeks as well as the number of trays outstanding for the year to date. This information is used to pinpoint excessive loss or unauthorized use of trays.

GLOSSARY OF LABELING AND BAR-CODE TERMINOLOGY

Following is a brief list of some of the terms used by salesmen and packaging engineers in describing the materials and processes used in labeling. The reader should be aware that these terms may have different meanings in other branches of engineering and technology.

acrylic-based adhesive – a mixture having acrylic polymers as the principal adhesive; effective over a fairly wide range of temperatures.

adhesive – a material causing the bonding of two surfaces.

aperture – in a scanner, the opening through which the light beam exits the unit. It determines the diameter of the light beam, and must be matched to the nominal bar size.

aspect ratio – the ratio of height to width of a bar code symbol.

background reflectance – measurement of the brightness or reflectance of the substrate on which a bar code is printed.

backing – the carrier sheet that covers the adhesive surface of pressure-sensitive label film.

bar – one of the darker, non-reflective stripes in a bar code.

bar code – an arrangement of vertical lines that provides computer translatable information based on the width and positioning of the lines according to standard systems.

bar width reduction – deliberate reduction of the width of bars on film masters or printing plates to compensate for the gain in bar width that takes place during the printing process.

basis weight – the weight in pounds of 500 sheets of paper cut to a specific length and width.

bidirectional code – a bar code that can be read in either direction across the bars and spaces.

butt cut labels – rectangular labels in continuous form, separated only by a cut across the strip (not by spaces).

caliper – label thickness in mils.

check digit – also, checksum character. A calculated number placed in a bar code to allow scanners to determine if they have read the code correctly.

clear area – an unprinted area left around a bar code to avoid interference by extraneous graphics (also called "quiet zone").

coating weight – weight in pounds of the adhesive on a specified area of the label.

code density – the number of characters that can be represented per unit length of space in a bar code.

continuous code – a bar code that does not have inter-character gaps as part of its structure.

contrast – difference of reflectance of the dark bars and the light spaces in a bar code.

converting – taking paper, foil, film, and other substrates, combining them and adding other materials, then printing, perforating, and coating to make labels, packaging materials, etc.

depth of field – distance between the nearest and farthest planes defining the limits of a scanner's accuracy when it is applied to a given printed example of a bar code. Does not apply to wand scanners since they must be placed directly in contact with the bar code.

die-cut – a cut made by the sharp edge of the die for the purpose of severing a label from the contiguous material.

edge error – irregularities in bar edges due to printing limitations.

EDP labels – pressure sensitive labels designed to be printed by electronic data processing equipment.

element – a single bar or space in a bar code.

emulsion side – the image-bearing side of a film master; the other side is called the base side.

encodation – characters (numbers and letters) available in a bar code.

face cut – a cut in the facesheet of the label (backing not cut).

facestock – paper, film, or foil stock used for making labels.

fade resistance – the relative capacity of a facestock to retain its colors or other appearance factors during exposure to light and other deteriorative influences.

fanfold – finished labels having perforated backing strips permitting the strips to be folded and stacked.

film master – a precision-made photographic image of a bar code symbol; used as a basis for making the printing plate.

first [time] read rate – The percentage of correct reads of a bar code on the first pass.

fixed beam scanner – a scanner using a beam of light that remains in one position and across which the bar code is passed.

flexographic printing – a kind of letterpress printing based on the use of fast drying inks and cylindrical ink transfer plates made of flexible plastics.

format – the physical or geometric arrangement specified for a particular bar code symbol.

guard bars – the three sets of two bars at the sides and center of UPC symbols. Also called "stop" and "start" bars.

hand scanning – use of a hand held "wand" or other manually movable sensor to read UPC symbols.

helium neon laser – a type of laser used in fixed-head barcode scanners; emits a red light of narrow spectral definition.

hot melt adhesive – an adhesive material that is solid at room temperature and is melted shortly before it is applied to the label.

inter-character gap – the spaces between the dark bars, or spaces representing different characters in a discrete character bar code.

laminate – a compound web (sheet, strip, etc.) formed by combining two or more webs of similar or dissimilar materials.

light pen – a hand-held scanning wand.

magnification factor – size of printed bar code compared to standard size.

mils – thousandths of an inch.

mis-encodation – using the wrong bar code to represent a character.

mis-read – when data output from a scanner does not agree with the actual bar code being read.

module – the smallest nominal width of bars and/or spaces in a bar code.

moving beam scanner – a bar code scanner that depends on a moving light beam to dynamically scan and decode a bar code symbol.

nominal bar element – width of a narrow bar or space.

no-read – or, non-scan: inability of a scanner to read a bar code due to improper orientation, excessive speed of scan, improperly printed bar code, or scanner/operator error.

numeric – a bar code that represents only numbers, not letters.

orientation – the positioning of graphic elements relative to a reference such as the base of a box, bottle, or can.

parity – system for encoding characters as "odd" or "even" bar code patterns. Used to provide a self-checking feature in bar codes.

perforations – series of small holes or slits, usually in straight lines, intended to help separation, feeding, or folding of the perforated material.

permanent adhesive – adhesive that gives long-lasting bond strength to a wide variety of surfaces.

pinfeed – a line of perforations at one or both sides of s strip of label stock that maintains the strip's alignment during the printing process.

pressure-sensitive – labels having an active adhesive layer that is protected by a silicone-coated release paper or film until the time of application.

printability range – range of print gain measured under actual working conditions on press-sheets chosen at random during a press run.

print contrast – comparison of reflectivity between bars and spaces.

print contrast signal (PCS) – a measurement of light reflectivity difference between the dark bars and the spaces in a bar code symbol.

print gain – the increase in bar width of a printed sample as compared to the size of the bar in the original artwork.

print quality – degree to which a bar code symbol complies with specifications for dimensional tolerances, edge roughness, spots, etc.

quiet zone – or, margin. A blank (unprinted) space before the start character or after the stop character.

reflectance – amount of light reflected back from a surface, measured under specified conditions.

removable adhesive – a pressure sensitive-adhesive providing a low level of adherence to the package.

repositionable adhesive – a pressure-sensitive material that allows repeated removal and application while retaining most of its sticking power.

resolution – sharpness, ability to reproduce fine detail; effectiveness of a scanner in reading narrow bars; dimension of the narrowest bar that can be read.

rubber-based adhesive – a pressure-sensitive adhesive containing substantial amounts of natural or synthetic rubber, often having special properties of importance in some applications.

scan – movement of the scanner's light beam across a bar code.

scanner – device that converts bar code symbols into electrical signals for data input and storage.

show-through – characteristic of a substrate permitting underlying markings and/or product to be seen; translucency, transparency.

space – lighter or unprinted areas separating the dark bars in a bar code.

start character – a pattern of bars and spaces used to identify the beginning of a bar code symbol.

stop character – a pattern of bars and spaces that identifies the beginning of a bar code symbol.

substrate – the film, paper, or other base material on which printing is done.

symbol contrast – difference between the largest and smallest reflectances in a scan profile.

tolerance – positive and negative deviations from the nominal value.

tooling – engraved cylindrical die used to make custom die-cuts or perforations.

truncation – decreasing the height of the bars in a bar code below the normally specified dimensions.

UCC – Uniform Code Council, an organization responsible for overseeing and administering the Universal Product Code.
UV varnish – a coating cured (solidified, "dried") with ultraviolet radiation; applied to label surfaces as a protective layer.

Version A – the standard version of the UPC bar code symbol
Version E – a special shortened version of the UPC, formed by using zero suppression.
void – unwanted white or light area in a bar caused by a printing error.

wide-to-narrow ratio – ratio of the width of the wide elements to the width of the narrow elements.

X-dimension – narrowest bar and space in a bar code.

zero suppression – technique used to shorten UPC symbols by omitting zeros from the bar code.

BIBLIOGRAPHY

ANON. 1989. Codes that can take it. Packaging Digest *26*, No. 3, 74, 76

ANON. 1993. Illuminating laser coding. Food Processing *54*, No. 5, 14.

FRASER, G. E. 1984. Imprinting, individual packages. Packaging Encyclopedia *29*, No. 4, 262, 265-266, 269.

KAYE, I. 1984. Adhesives, cold water borne. Packaging Encyclopedia *29*, No. 4, 53-54.

LINGLE, R. 1988. Showcase: Non-contact coding. Prepared Foods *157*, No. 7, 103-106.

MORRIS, C.E. 1988. Monitoring fresh-food life with T/T labels. Food Engineering *60*, No. 4, 52, 54, 57.

QUINN, T. 1984. Adhesives, hot melt. Food Processing *29*, No. 4,

RICE, J. 1993. Consumers call for open-dating. Food Processing *54*, No. 4, 101-103.

RICHTER, R.T. 1984

RILEY, R. W. 1984. Labeling, pressure sensitive. Packaging Encyclopedia *29*, No. 4, 272, 275-276.

ROOKS, A. 1990. The 21st century UPC: Same design, different results. *159*, No. 1, 34.

SACCHAROW, S. 1985. Label functions. Food Packaging and Labeling *9*, No. 12, 1-6.

SCHIESSEL, H. 1984. Labeling, glue and thermoplastic. Packaging Encyclopedia *29*, No. 4, 269-272.

SUTTON, J. 1992. Which device is best for bar coding labels? In-Plant Reproductions, July 1992, p. 25.

chapter ten

REGULATIONS AFFECTING PRODUCT DEVELOPMENT

INTRODUCTION

As the product developer works at his primary task of meeting the requirements of the public and his employer, he must not neglect the need to conform to the hodgepodge of federal, state, and local laws that regulate virtually every aspect of the designing of new products and packages. Some of these regulations have been referred to in other chapters where their consideration was important in relation to the subject then being discussed. In this chapter, most of the attention will be concentrated on the Nutrition Labeling and Education Act and its subsequent modifications.

In 1990, Congress enacted a law called the Nutrition Labeling and Education Act, the ostensible purpose of which was to instruct the FDA to draw up new, modified, uniform, and (of course) more stringent rules on label claims for foods. As a consequence, FDA proposed, in 1991, new rules regulating label claims for nutritional content, health benefits, serving sizes, etc. Time was allowed for comments from industry and other interested parties, the revised and final regulations being scheduled for release in November 1992.

The deadline was not met, but the FDA announced in late November, that, by provisions of the law, the proposed rules would go into effect anyway. The agency also announced that it was preparing to publish regulations to supersede the laws automatically implemented in November. A small difficulty for food manufacturers and other packagers was that there was no published version these firms could obtain for guidance in making their products conform to the rules.

Eventually, a semi-final document was assembled and published as the Nutrition Labeling and Education Act of 1993 (NLEA); it occupied much of the two volumes of the Federal Register dated January 6, 1993. This complex set of regulations includes foods controlled by both the FDA and the USDA, the latter's interests being concentrated on processed meat and poultry products. General compliance was postponed until May 1994, a relatively short time for manufacturers who must redesign labels, perhaps have hundeds of analyses conducted and the results reviewed and verified, and then have conforming labels printed and delivered to the packaging lines.

Up to the time this article was written, the most extensive changes following the January 1993 compendium were included in the August 18, 1993 issue of the Federal Register. It can be confidently predicted that there will be more changes, many more. In the August publication, the FDA also included 53 pages of answers to the questions most frequently asked about the NLEA. Litigation by industry to modify or abrogate some of these regulations, or all of them, will doubtless become part of the picture.

In the following section, attempts will be made to summarize and explain those NLEA provisions that are of the greatest important to persons engaged in some part of the food product development process. Of course, condensing several hundred pages of legalese to a chapter of about 20 pages will undoubtedly leave a lot of gaps. Consult the original publications for all the original lyrical prose. The Federal Register issues of Jan. 6, 1993 are available on computer floppies (Superintendent of Documents, U.S. Government Printing Office, Washington. DC; phone your orders to 1-202-512-1530 or FAX them to 1-202-512-0132, then wait and hope; price was $88.50 for four diskettes).

GENERAL LABELING REQUIREMENTS

For many years, there have been laws and regulations specifying the format and content of labels, all of them having the implicit or expressed intent of protecting and informing the consumer. Not all of these statutes refer directly to nutritional claims, some of them are concerned with economic details or other matters thought to be important to the purchaser of the goods. Some of the non-nutritional elements of food labels that have been the subject of federal legislation and FDA actions are definitions of principal and information panels, instructions for listing ingredients, and weight declaration formats. The most important of these non-nutritional labeling requirements will be covered under this heading.

Principal Display Panel and Information Panel

The principal display panel is that portion of the package label most likely to be displayed, presented, shown, or examined under customary conditions of display for retail. It must contain, as a minimum, the statement of identity for the food product and the net quantity of contents in avoirdupois and metric units.

The information panel is that part of the label immediately contiguous to, and to the right of, the principal display panel as the latter will be seen when an observer faces the principal display panel. Mandatory information that must be grouped together on the information panel includes a declaration of ingredients, the name and address of the maufacturer, packer, or distributor of the food product, and the nutrition labeling material. The information panel must not be separated from the principal display panel by any intervening material, such as ad copy, bar codes, recipes, coupons, etc., and such material must not separate the mandatory information in the information panel.

Ingredient Declarations

The FDA has struggled for decades over the proper nomenclature for ingredients and the order in which to list them. Regulations in effect at the time of this writing will be given. Experience indicates that sporadic changes will continue to occur in this regulatory category.

Order of predominance. Generally, the ingredient declaration (ID) must state the common or usual name of each ingredient in the product in descending order of predominance (i.e., quantity). If an ingredient having a common or usual name contains several components, the compound ingredient may itself be listed followed (within parentheses) by its list of ingredients in descending order of quantity; or, each of the components (ingredients) in the compound ingredient may be distributed throughout the main ID in the order of their predominance in the finished product with no mention of the name of the compounded ingredient.

Those ingredients each of which constitutes less than 2% of the finished product do not need to be listed in order of their predominance if they are preceded by the statement, "Contains 2% or less of. . ." An FDA representative (Campbell 1993) said, "The general provision is that ingredients are listed in descending order of predominance by weight with the provision that ingredients at 2% or less can be listed in a group, but not in [any particular] order."

Spices, flavorings, and non-certified food colors may be labeled as such. The protein source used to make hydrolysates (for flavorings) must be given. Casein must be declared as a dairy ingredient in nondairy foods. Listings of mono-sodium glutamate, sulfites, and FD&C Yellow No. 5 must include a warning statement.

Juice content. The NLEA included provisions on declarations in the ingredient listing and a requirement for declaration of percent of juice in a beverage, the effective date later changed to May 8, 1994. Additionally, there were many regulations affecting the way components were described or the methods used for calculating their amount.

Added food colorants. Some color additives must be declared by name. Previously, most of the ingredients used as colors could be combined under the words "color" or "coloring." The new rules require FD&C colors to be declared by name: FD&C Blue No. 1, FD&C Blue No. 2, FD&C Green No. 3, FD&C Red No. 3. FD&C Red No. 40, FD&C Yellow No. 5, and FD&C Yellow No. 6. There is, however, a provision allowing abbreviations for the colors, e.g., "Blue 1" for "FD&C Blue No. 1." Non-certified colors can still be declared under the generic term,"color."

Standardized Foods. The other major ingredient list change is in the ingredient listing for standardized foods. Previously, some of the ingedients in foods for which there was a standard of identity did not have to be declared. Standardized foods are now subject to the same ingredient declaration requirements as all other foods — each ingredient used in the food must be declared by name.

Beta-carotene. This substance can be reported in 2%, 5%, and 10% increments, the same as the other vitamins and minerals. Since beta-carotene is a "sub-component" of vitamin A, it must comply with the regulations for reporting sub-components, i.e., it must either be listed indented under the vitamin A declaration in column format or be enclosed in parentheses if reported beside vitamin A in the horizontal type of listing.

Fats. Since there would ordinarily be no fatty acids in foods that are "fat free," the former requirement for reporting both mono- and poly-unsaturated fatty acids on the labels of these foods (would show zero in all cases) has been deleted. Low-fat and higher fat foods must still bear label statements showing their content of these fatty acids. Saturated fatty acids are still to be reported on all foods, presumably including those with no fat content.

The "and/or" conjunction much used in reporting fats and oils used as ingredients (e.g., "cottonseed and/or soybean oil") mandates declaration of saturated fat content of the most saturated fat listed; in many cases this will be coconut oil, or palm oil, or beef fat. Presumably, hydrogenation (leading to increased saturation) will have to be taken into account.

Miscellaneous Ingredients. Rules on ingredients will require the declaration of all protein hydrolysates (all types), including identification of the food source, e.g., "hydrolyzed corn protein," in the list of ingredients. Also, there will be required the description of caseinates as milk derivatives when these materials are used in foods that claim to be nondairy foods. Also, the declaration in standardized foods of sulfiting agents having a functional effect or that are present at a detectable level of 10 or more parts per million. Certain exemptions from requirements are provided for listing the specific common or usual names of waxes and coatings on fresh fruits and vegetables. There are also provisions for optional declarations, such as a format for voluntary declaration of percentage ingredients.

Metric Declaration of Contents

The metric amendment to the Fair Labeling and Packaging Act was scheduled to become effective on February 14, 1994, according to current law. It will require the net contents statement on all product labels to appear in both avoirdupois and metric units. The government has announced, however, that it will not take any enforcement action on this regulation until the new nutritional labeling law (FDA part of it) takes effect, supposedly on May 8, 1994. It is noteworthy that a space must be left between digits and metric symbols (the short forms are "symbols" not "abbreviations"), as pointed out by Antoine (1994).

NUTRITION LABELING

The NLEA forces many changes in the practices of nutrition labeling. Nutrition labeling now is required for almost all processed foods that are meaningful sources of nutrients. The safest approach for the manufacturer is to assume that *all* materials used as, or usable as, foods come within the purview of these regulations. Very detailed specifications for format, terminology, content, and methods of calculation were promulgated. It is not possible to cover all of these details in the present discussion, but some attention will be given to the important points.

Format of the Nutrition Fact Panel

Food labels are now to include a revamped "Nutrition Fact Panel" (NFP) that must appear on the information panel and be enclosed in a box (frame) demarked by a thin line. It will be headed with a new title, "Nutrition Facts." There is no specific requirement of minimum or maximum size for this panel, although, as will be seen from later discussions, a minimum size is effectively dictated by other provisions. The panel should be placed in an position that allows it to be easily read. The printing should be in black or one-color type on a white or neutral contrasting background whenever practical, but this does not specifically prohibit reverse printing (such as white on black) or the use of other colors. Consumer readability is of the essence, however.

Type, ruling, and other features must be consistent throughout the area within the box, that is, highlighting or emphasizing a particular statement or nutrient amount through use of a different color, type font, or the like is not acceptable. The FDA "urges that the panel is readily observable and legible to consumers at the point of purchase . . ."

Information to be presented in the NFP must disclose the amount of each food or food component present in a standard serving size of the food product as well as the number of servings contained in the package.

Size and shape of the panel. The regulated nutrition display panel must be set off in a box shape with 90° corners, according to all examples shown in the Federal Register. This indicates that oblong, oval, round, or other non-rectangular shapes are unacceptable regardless of the shape of the package or esthetics of the graphics as interpreted by the advertising agency.

Letters, words, symbols, or decorative material other than described in the regulations as mandatory or optional must not appear within the NFP.

Size of the panel and some other dimensions are based on a percentage or fraction of the total surface available to bear labeling. A uniform method of determining the total surface is therefore essential. The Federal Register states, "If it is technologically feasible and practicable to apply a label to the surface area, and it is an area likely to be seen by the consumer under normal handling conditions, it should be considered as an area available to bear labeling."

It is not necessary to include in the available surface area the package flanges; tops and bottoms of cans; shoulders, necks, and caps of bottles and jars; folded flaps; and other normally unused areas unless you are presently using these areas for labeling purposes. Windows of cellophane or other transparent films are considered available space if you are now printing on them, otherwise not. Current non-mandatory copy (including bar codes) on the package does not have priority – it must be counted as part of the total surface area and, when necessary, it must be removed to leave a place for the required information.

If the package has less than 12 square inches of space available to bear labeling, nutrition labeling is not required but the manufacturer's name and address (or phone number) must be printed as a source of nutritional and other information. There are two options for packages with 12 to 40 square inches available for labeling: (1) A horizontally formatted tabular display using many abbreviations and no footnotes; and (2) A "minimum" nutrition panel that has all of the nutrition label information in the same format up to and including the declaration of "iron."

Orientation of the panel. The preferred, or standard, arrangement of the Nutrition Facts information appears to be a single vertical column, but there are two other formats that may be used depending upon the size of the package and certain other considerations; these are the tabular format, in which two or more columns of information are presented, and the linear, or so-called string, format that presents the information with each nutrient declaration followed immediately on the same line by another statement and so on until the end of the line is reached, then continuing on the next line with more data until the complete nutritional information requirement has been presented.

Use of the tabular format is justified only when the food qualifies for the simplified format (qualifications for the simplified format are described elsewhere). and the available space is less than 40 square inches or has an insufficient vertical dimension to accommodate either the full format or the footnote-to-the-side format.

If there is not sufficient continuous vertical space (about three inches) to accommodate the required components of the nutrition panel up to and including the mandatory declaration of iron, the nutrition label may be presented in a tabular display; in the interests of consistency, macronutrients may be split into more than two columns. The listing of any additional vitamins and minerals other than those required may appear horizontally below the list of mandatory vitamins and minerals or in a continuation immediately above the Daily Value footnote to the right. If the phrase, "Not a significant source of [nutrient]" is required, it must appear under a bar separating it from the nutrients declared above it.

The nutrition facts panel can be printed on a sticker or label and adhered to the package (Anon. 1994). This is an option that is of considerable advantage to manufacturers/packagers that make short runs of several different varieties, all packed in about the same kind of package and labeled by the packer with stickers printed in its own plant, as required, by electronic composing devices. Bakers making several kinds of cakes, cookies, bread, etc., every day should find this to be a practical and relatively economical alternative.

Type font and spacing. In the January 1993 publication, the FDA went into considerable detail on specifications for the type to be used in the printing appearing in the Nutrition Facts panel, with the laudable goal of providing the consumer with readable information. For example, the regulations specified the amount of space that must appear between adjacent letters in a word. Since the practical, esthetically preferable, or standard measurement for this "kerning" will clearly vary depending upon the type style or font used, there could be hundreds of possible variations. The requirement was later changed to state that sufficient space must appear between the letters so that they do not touch.

The type font specified was Helvetica (only), but now any legible type style can be used. All words must be formed of upper and lower case letters. Core nutrients are to be printed in at least 8-point type while footnotes have to be 6-point size or larger. The "Nutrition Facts" heading must be in the largest type face of any other wording in the panel, i.e., at least a point size larger than the size used for printing the core nutrients. The thickness of the bars or ruling that separate different parts of the panel is not specified.

In the simplified formats, the "Beta-carotene" amount and the "Not a significant source of:" statement must appear in 8-point size or larger. The "Nutrition Facts" heading must be printed in a larger type size than any of the other information in the NF box but no smaller than 8-point.

For those nutrients listed at the option of the manufacturer, i.e, "voluntary nutrients," "Potassium" should appear in bold-faced type directly under "Sodium." Potassium must not be indented or be placed in parentheses. Other vitamins and minerals optionally declared must follow "Iron" in the order: Vitamin D, Vitamin E, Thiamin, Riboflavin, Niacin, Vitamin B_6, Foliate, Vitamin B_{12}, Biotin, Pantothenic Acid, Phosphorus, Iodine, Magnesium, Zinc, and Copper.

Simplified Label Format

Certain foods may be entitled to use a simplified label arrangement. This simplification will be allowed when the food contains insignificant amounts of seven or more of the mandatorily-reportable nutrients. Some carbonated beverages are examples of foods on which such stripped-down wording is allowed. When the simplified format is used, information on total calories, total fat, total carbohydrate, protein, and sodium must be listed, even though they are present in insignificant amounts. Other nutrients that may be present, along with calories from fat, must be declared. The phrase, "Not a significant source of the following . . ." is required only if: vitamins and minerals are required to be added to a product for which a Standard of Identity exists, or vitamins or minerals are voluntarily added to a food, or voluntary nutrients occur naturally in more than insignificant amounts, or nutrient content claims are made on the label of the product. Alternative label formats may also be used if the available label space is not adaptable to the standard requirements.

Declarations Required Within the Panel

A typical NFP will include statements of the serving size and servings per container directly under the heading "Nutrition Facts." Determining these two declarations (the second being a derivative of the first, of course) have the possibility of causing more confusion than many of the other provisions in this law. Below are some guidelines that may help.

Serving size. Serving size is to be expressed in common household units (cups, spoonfuls, etc.) followed by the metric equivalents in grams or milliliters that approximate the Reference Amount (RA). "Common household measurements" have been defined as: Teaspoon (tsp) = 5 ml; tablespoon (tbsp) = 15 ml; cup = 240 ml; fluid ounce (fl oz) = 30 ml; and ounce in weight (oz) = 28 g. Sizes may be expressed in one-fourth or one-third increments for cups, in one-fourth increments for teaspoons, and in one-half ounce increments for weight.

Both the FDA and the USDA have established so-called Reference Amounts (RA) for a wide variety of foods based on an amount that is "customarily" consumed per "eating occasion" by a member of the US population more than four years of age. A serving size for each labeled product is to be established in whole units most closely approximating its specific Reference Amount. The following rules are to be used in establishing a serving size:

(1) For products sold in discrete units, such as slices of bread (or individual packets within a multi-serving package, e.g., several candy bars in an overpack) serving size shall be the number of units that most closely approximates the reference amount (RA), within the range of 67% to 200% of the reference amount. An amount over 200% of the RA may be labeled as

a "serving" if the whole unit could reasonably be expected to be consumed in a single eating occasion. In addition, if the unit weighs from 50% to 67% of the RA, the producer may declare one unit as the serving size.

(2) For those items sold in large separate units, such as pies, cheesecakes, etc., intended to be used for serving several persons (or several times to one person) the product may be divided so that a piece most closely approximates the RA for the product category; a fractional amount must be divisible by 2 or 3.

Certain items have special rules, a few of which will be given here.

(1) Serving sizes for single serving meals, main dishes, and the like will be the entire edible content of the container.

(2) A variety pack, such as a package containing several varieties of foods, each being a single serving, shall provide nutrition information for each variety of food on the basis of the serving size derived from an appropriate RA.

(3) The label on a package containing one plus a fraction servings would have to declare the whole amount as a single serving.

(4) For mixes and the like that make multiple portions when fully prepared, the RA is the amount required to make one standard serving.

(5) If the product consists of pieces packed in liquid that is not ordinarily consumed (e.g., dill pickles), drained weight will be used as the serving size and for nutrient testing.

Products intended to be used for weight control and available only through programs for weight-control or weight-maintenance may have their serving sizes determined by the manufacturer so as to be consistent with the goals of the program. But, these products must bear the statement, "For sale only through the . . . program." Nutrient content and/or health claims must be based on the RA of the product.

The RA for an imitation or substitute food, or for an altered food such as a low calorie version, shall be the same as the food for it is intended to replace. If a food with an RA in grams is modified by aerating it, the bulk density being lowered by 25% or more, the manufacturer may determine the RA by adjusting for the difference in density, with rounding in 5 g increments; the product of decreased density must be described as "whipped" or "aerated."

A few of the FDA-specified RAs are included below to show the general pattern (LS = label statement):

(1) Biscuits, croissants, bagels, tortillas, soft bread sticks, soft pretzels, corn bread, hush puppies: RA = 55g; LS = number of pieces.

(2) Pies, cobblers, turnovers, fruit crisps, other pastries: RA = 125 g; LS = number of slices or fraction of the whole pie.

(3) Coffee cakes, crumb cakes, donuts, Danish, sweet rolls, quick breads, muffins, and toaster pastries: RA = 55 g; LS = number of pieces or fraction of the cake.

(4) Crackers (but not snack crackers), melba toast, crisp bread sticks,

cones: RA = 15 g; LS = number of pieces.

(5) Hot breakfast cereals, hominy grits: RA = 1 cup as made, 40 g dry, 55g flavored or sweetened; LS = number of cups.

Servings per container. Servings per container are to be expressed in whole numbers. For products containing from 2 to 5 servings per container, the number of servings may be expressed in ½ serving increments, if appropriate. Use of the terms "approximately" or "about" are permitted as designations of the number of servings if the number is not a whole number.

Information on the amount of nutrients in each serving will be in accordance with the servings-per-container and the serving-size determination. This includes the amount of each of the designated nutrients in grams or milligrams along with the Percent of Daily Value (PDV) for each. The PDV will be expressed in 1% increments and is based on the amount of the nutrient present in the serving divided by the amount of that particular nutrient that comprises the reference daily diet of 2,000 calories. These reference amounts appear further down in the label under the 2,000 calories heading.

Daily Reference Values have been calculated on the basis of 2,000 calories as follows: 30% of calories as fat, 10% of calories as saturated fat, and 10% of calories as protein. Other DRVs are total fat less than 65g, saturated fat less than 20 g, cholesterol less than 300 mg, and sodium less than 2,400 mg.

Abbreviations for nutrients. In those situations where abbreviations are allowed (see qualifications elsewhere) in the Nutrition Facts box, the following standard forms must be used:

Serving Size	Serv size
Servings per container	Serv
Calories from fat	Fat cal
Calories from saturated fat	Sat fat cal
Saturated fat	Sat fat
Polyunsaturated fat	Polyunsat fat
Cholesterol	Cholest
Total carbohydrate	Total carb
Dietary fiber	Fiber
Soluble fiber	Sol fiber
Insoluble fiber	Insol fiber
Sugar alcohol	Sugar alc
Other carbohydrates	Other carb

Mandatory and optional listings. The new regulations changed the number of vitamins, minerals, calories, and other factors that must, or can, be included in the dietary components declared in the nutrition panel. The components and the order in which they are to appear are shown below.

The mandatory components are designated by an asterisk, all others are optional and appear at the packager's discretion or as triggered by some other provision.

total calories*
calories from fat*
calories from saturated fat
total fat*
saturated fat*
stearic acid [meat and poultry products only]
polyunsaturated fat
monounsaturated fat
cholesterol*
sodium*
potassium
total carbohydrate*
dietary fiber*
soluble fiber
insoluble fiber
sugars*
sugar alcohols
other carbohydrates
protein*
vitamin A*
vitamin C*
calcium*
iron*
other essential vitamins and minerals.

Content information on nutritional factors not listed as either required or optional declarations (such things as omega-3 fatty acids or boron) may be printed on the label provided the statement is true and not misleading. It cannot appear in the Nutrition Facts box, however, and it must be limited to statements of amount and not declare "High in boron" or make other comparative claims. Any such statements will trigger the need for full nutritional labeling.

Reference Daily Intakes

The revised food label includes the new basic standard, "Reference Daily Intakes (RDIs)," to replace the term "U.S. RDA," which was introduced in 1973 as a label reference value for vitamins, minerals, and protein in the voluntary nutrition labeling format. The name change is supposed to eliminate the confusion that was said to exist about the meaning of the term "U.S. RDA" (values derived by FDA for use on food labels) and "RDA" (values published by the National Academy of Sciences for various population groups and used by FDA to determine U.S. RDAs). At least for a time, the RDI values will, in fact, be the same as the old U.S. RDAs. The FDA is said to be considering the setting of new RDI values to "reflect the current state of nutrition science."

Under provisions of the Dietary Supplement Act of 1992, the FDA was prohibited from making changes in the RDIs until November 9, 1993. As of this writing, no changes have been proposed but they are certainly likely to occur.

At the present time, the following Reference Daily Intake values are in effect:

Vitamin A	5000 IU
Vitamin C	60 mg
Thiamine	1.5 mg
Riboflavin	1.7 mg
Niacin	20 mg
Calcium	1000 mg
Iron	18 mg
Vitamin D	400 IU
Vitamin E	30 IU
Vitamin B_6	2.0 mg
Folic acid	0.4 mg
Vitamin B_{12}	6 micrograms
Phosphorus	1000 mg
Iodine	150 micrograms
Magnesium	400 mg
Zinc	15 mg
Copper	2 mg
Biotin	0.3 mg
Pantothenic acid	10 mg

Nutrient Content per Serving

The format for declaring nutrient content per serving was revised. The new regulations state that all nutrients must be declared as percentages of their Daily Value (the new label reference amount). The content, in metric units, of macronutrients such as fat, cholesterol, sodium, carbohydrates, and protein, still must be placed to the immediate right of the name of each of these nutrients. In addition, a new column, to be headed "% Daily Value," will replace "U.S." in the statement.

The content of micronutrients (vitamins A & C, optional nutrients, calcium, iron, and optional minerals) will continue to be expressed as a percentage of current U.S. reference values, but the term "Daily Value" will replace "U.S."

Calorie calculations. In the January publication, it appeared that five alternative methods for calorie calculations were permissible: (1)

specific Atwater values for different foods and food components, as found in tables of these values; (2) the common or usual method, which assumes fixed values of 4 calories for each gram of protein and carbohydrate and 9 calories for each gram of fat; (3) the 4, 4, 9 system with insoluble fiber included in the total carbohydrate; (4) use of special data for certain food ingredients such as maltodextrins; or (5) bomb calorimetry determinations conducted on the finished food.

As most food product developers know, it is much more convenient to use the approximate values of 4 calories for a gram of carbohydrate or protein and 9 calories for a gram of fat, than to use the specific Atwood factors. The latter method leads to problems in correlating your material with a name listed in the Atwood tables, among other difficulties. Bomb calorimetry is seldom used in the average industrial food laboratory; small companies will not even have the equipment for this test.

The August 18, 1993 publication disclosed that calorie declarations should be calculated with as much precision as possible up to the point of final rounding so as to provide consumers with the most accurate information on the caloric content of a serving of the food. Where specific or general food factors are used, the factors should be applied to the actual amounts, not the rounded amounts, of protein, fat, and carbohydrate present in a serving.

Use of databases. Many manufacturers use databases instead of specific analyses to obtain the figures that they enter in nutritional statements. For example, if the product contains 50% flour, they would consult one of the reference sources in which the composition information they needed could be found, and, if the reference states that "flour" contains 11% protein, they would assume that 0.11 times 0.50 times the portion weight gave the amount of protein (from flour) in a serving, and add this figure to protein contributions from other ingredients (obtained in the same way) to determine the protein content claimed on the product label (allowing for rounding, etc.). Among the databases that have been widely used is the Agricultural Handbook 8 series, the twenty-odd volumes of which (some still in preparation) contain analytical values for over three thousand raw materials, ingredients, and foods. The present and prospective titles in this series are:

8-1. Dairy and egg products.
8-2. Spices and herbs.
8-3. Baby foods.
8-4. Fats and oils.
8-5. Poultry products.
8-6. Soups, sauces, and gravies.
8-7. Sausages and luncheon meats.
8-8. Breakfast cereals.
8-9. Fruits and fruit juices.

8-10. Pork products.
8-11. Vegetables and vegetable products.
8-12. Nut and seed products.
8-13. Beef products.
8-14. Beverages.
8-15. Finfish and shellfish products.
8-16. Legumes and legume products.
8-17. Lamb, veal, and game.
8-18. Baked products.
8-19. Snacks and sweets.
8-20. Cereal grains and pastas.
8-21. Fast foods.
8-22. Mixed dishes.

Nutrient data bases prepared by the Human Nutrition Information Service of the U.S. Department of Agriculture are available for use by the food industry and others. The basic data set is the USDA Nutrient Data Base for Standard Reference (Anon. 1990). It contains all nutrient values published in Agriculture Handbook No. 8 and is updated continually.

A collection of sources for nutrient data, not seen by the author, can be found in the "Nutrient Data Bank Directory" (Smith 1992). This, used with information from the "Journal of Dietetic Software," is said to provide valuable guidance that will assist the researcher in selecting and using an appropriate nutrient data base system.

Shapiro (1993) states that the FDA position, as put forth in the "FDA Manual: A Guide for Developing and Using Data Bases"(1992 Edition) does not indicate approval for use of databases for calculating nutrient values: i.e., "FDA is not accepting ingredient composition data bases to calculate the final composition of a mixed product." Further, the FDA questions the reliability of the databases currently available, ". . . some existing databases have no documented history pertaining to their origin. Consequently, the quality of the data for labeling purposes cannot be assessed." Standard AOAC procedures (or other approved analytical methods) will be used by the FDA to determine compliance. There always exists the possibility that the FDA will approve databases for use in calculating nutrient claims. In fact, it appears that a database developed by the American Institute of Baking has been accepted by the FDA. Wilson Foods recently announced it had received USDA approval for using the Genesis database (Orloski 1994).

Software for facilitating compliance with labeling regulations has begun to appear. A software package described by Sperber (1992) is said to include all USDA label regulations and sample label formats as well as other USDA forms. An FDA publication (Troiano 1994) lists the following resources (possibly of greatest interest to consumers): "The New Food Label and You," available free from USDA-FSIS Room 1180-S, Washington DC 2050 and "Food Labels," a 6-minute video that can be obtained from FDA Office of Public Affairs, 5600 Fishers lane, HFI-40, Rockville MD 20857.

Nutrient data by analyses. Another way of determining the quantity of (e.g) protein that must be declared in the nutritional statement, is to subject samples of the finished product to official analytical procedures for determining protein, apply the appropriate factors to obtain the amount of protein in a serving, and place this information (rounded, etc., as required) in the nutrient panel. It is very obvious that it is faster and much less costly to use databases than specific analyses for this purpose. It is probable, though not necessarily certain, that analyses performed on the product would give more accurate figures than would the database calculations. Whether the differences would be of practical importance to any consumer is debatable.

Claims of Special Nutritional Value

As is well known, food marketers like to apply to their packages descriptions of dietary values or physiological effects that are thought to increase the consumer-appeal of the products: "light," "healthy," "natural," "organic," "wholesome," and "nutritional" are examples of popular selling words. The NLEA proclaimed official definitions of some of these terms and confessed temporary defeat in defining others. They called these words and phrases, or at least some of them, "nutrient content descriptors."

Nutrient content descriptors. The NLEA established rules governing the use of nutrient content descriptors in label claims purporting to describe the level of a nutritional factor in a food and defined some of these terms. The "core" terms include "free," "low," "high," "reduced," "less," "light," and "more." The 1993 regulations say that nutrient content descriptors not defined by the FDA cannot be used in food labeling. Official definitions of terms acceptable under some conditions are:

Free – the product must be absolutely free of, or contain "physiologically inconsequential amounts" of, a nutrient. The term can be used in conjunction with the following components: fat, saturated fat, cholesterol, sodium, sugars, and calcium. For example, "calorie-free" means that fewer than five calories are present in a serving, and "sugar-free" and "fat-free" both mean less than 0.5 grams of sugar or fat, respectively, are present in a serving. Terms that may be used instead of "free" are: "without," "trivial source of," "dietarily insignificant source of," "negligible source of," "no," and "zero."

Lean and extra-lean – the FDA adopted USDA's definition for these terms to describe the fat content of seafood and game meats. Lean meat has been defined as having no more than 10 grams of fat, 4 g saturated fat, and 95 milligrams of cholesterol per serving and per 100 grams. Extra lean has been defined as having five grams of saturated fat and 95 mg of cholesterol per serving or per 100 grams.

This definition was revised in the August 1993 publication to provide that products containing less than 4.5 grams of saturated fat per serving or per 100 grams will be entitled to use the description, "lean." Also, the rounding rules were changed so that foods containing less than five grams per serving must have the amount declared in ½ gram increments. Foods containing more than five grams must declare these nutrients in full one-gram increments. There is a new definition of fat as the sum of all lipid fatty acids expressed as triglycerides – this has the effect of eliminating all types of fat analyses except a chromatographic fatty acid profile (Hall 1993).

Good source – indicates that one serving of the labeled food contains 10% to 19% of the Daily Value for a particular nutrient.

High – this term can be used only if the food contains in each serving 20% or more of the Daily Value for a particular nutrient.

Reduced – (or "less" or "fewer") means that a nutritionally altered foodstuff contains 25% or less of a nutrient, as compared to the reference (unaltered) product. This claim cannot be made if the reference food already meets the requirements for a claim of "low."

Low – describes a food (altered or not) that contains 25% less of a nutrient than the reference material. For example, pretzels that contain 25% less fat than potato chips could carry a "less" claim.

Light – (or "lite") this descriptor, if used without any qualifier, means that the nutritionally altered product contains one-third fewer calories or half the fat of the reference food. If the food derives more than 50% of its calories from fat, the reduction is required to be 50% of the fat.

When the sodium content of a food has been reduced by 50%, "light in sodium" may be used as a descriptor regardless of the fat content. If foods already describable as low-fat and low-calorie have their sodium content reduced by 50%, the term "light" can be applied.

The term "light" can still be used to describe properties such as texture and color, as long as the label explains the intent, for example, "light brown sugar."

More – (or "added") means that a serving of either an altered or unaltered food contains a nutrient at a level of at least 10% more of the DV than the reference food. The 10% of Daily Value would also apply to "fortified," "enriched," and "added" claims, but in those cases the food must be altered.

"Healthy". The FDA decided that "healthy" should be defined in a way that restricts the contents of those nutrients with public health significance to levels that do not exceed the disclosure level for the nutrients, and should focus on restrictions for fat and saturated fat. It was proposed that "healthy" be allowed to describe foods that meet FDA definitions for low fat and saturated fat and in which neither cholesterol or sodium are present at levels exceeding the appropriate disclosure levels.

Other "Health" claims. A health claim is a label statement referring to the relationship between a food or nutrient and the risk [probability?] of a disease or health-related condition. Such claims are prohibited by the

NLEA, unless they are made in accord with FDA regulations. The law also instructed the FDA to evaluate ten nutrient/disease relationships to determine whether claims should be permitted.

Under the final regulations, health claims were allowed for seven relationships between a food or nutrient and the risk of disease or health-related conditon. The claims must meet certain requirements – they must be phrased so that the consumer can understand the relationship between the nutrient and the disease and the nutrient's importance in the relationship within the context of the total daily diet. The seven approved health claims are: (1) osteoporosis and calcium; (2) coronary heart disease (CHD) and saturated fat and cholesterol; (3) cancer and fat; (4) cancer and fiber-containing grain products, fruits, and vegetables; (5) risk of CHD and fiber-containing foods (specifically, fruits, vegetables, and grain products); (6) hypertension and sodium; and (7) cancer and fruit, vegetables, foods that are low in fat and may contain vitamin A.

The claims for food:disease relationships that the FDA did not accept included: (1) cancer and dietary fiber; (2) cardiovascular disease and dietary fiber; (3) cancer and antioxidants; (4) cancer and vitamins; (5) immune function in the elderly and zinc; and (6) neural tube defects and folic acid. However, in a very recent publication (Fed. Reg. Oct. 14, 1993), the FDA proposed to: (1) require folic acid fortification of grain products at specified levels to help women of childbearing age take in at least 0.4 mg folate per day to prevent neural tube defects; (2) restrict folic acid addition to breakfast cereals, standardized foods requiring it, infant formulas, dietary supplements, and foods for special dietary purposes; and (3) allow foods and dietary supplements to bear a label claim about the relationship between folate and lowered risk of neural tube defects.

Natural. The word "natural" is on every food marketer's lips and will be on his labels, if at all possible. A representative of the FDA made the following statement (Campbell 1993): "The status of the word 'natural' has not changed. We did not establish a regulation on 'natural' in these final regulations. We had attempted to define 'natural,' but did not. So, it will be regulated under the general provisions that say that nothing on the label that may be misleading."

From this statement we can conclude that, if you are using the adjective "natural" on your food label, your legal position remains the same as before, whatever that was. It was not changed by the NLEA. The FDA, or their superiors, seemingly could not reach a consensus on a proper definition or usage of this word when the regulations were being developed. "Natural" will be continue to be regulated under the general provisions that say nothing on the label may be misleading.

Meals and Main Dishes

The FDA promulgated definitions or criteria for packages containing "meal type" and "main dish type" products.

A "meal" product is defined as a food that is represented as, or is commonly understood to be a breakfast, lunch, dinner, or meal, and makes a major contribution to the consumer's total diet by:

(1) Weighing at least 10 oz per labeled serving.

(2) Containing not less than 40 g each of at least three different foods representing two or more of the following four food groups (none of them being gravies, relishes, condiments, pickles, jams, jellies, syrups, breading, garnishes, or sauces other than dairy-based sauces):

(a) Bread, cereal, rice, or pasta group.

(b) Fruits and vegetables group.

(c) Milk, yoghurt, and cheese group.

(d) Meat, poultry, fish, dry beans, eggs, and nuts group.

A "main dish" product has a similar definition except that (1) it must weigh 6 oz per labeled serving, and (2) it must contain not less than 40 g of at least two different foods from two or more of the food groups listed above.

Miscellaneous Rules

The rules governing claims of nutrient content also contain a definition for percent fat free, provisions for implied claims, and provisions for use of descriptor terms on meals and main dishes. There are several requirements for additional label statements when nutrient content claims are made. The NLEA requires a statement on the main panel of a label saying, "See (side) panel for nutrition information."

When another nutrient is present in the food "at a level which increases to persons in the general population the risk of a disease or health-related condition which is diet related," the name of that nutrient must be included in this referral statement (for example, "See side panel for information on sodium and other nutrients").

Listing of single amino acids or maltodextrin, and separate listing of calories from polyunsaturated fats and calories from carbohydrates may not appear as part of "Nutrition Facts."

For percent Daily Value calculations, either rounded or unrounded analytical values can be used depending upon which one leads to the most consistent label, i.e., when percent DV and nutrient content are compared.

The regulations provide that: "Reasonable excesses of a vitamin, mineral, protein, total carbohydrate, dietery fiber, other carbohydrate, poly- or mono-unsaturated fat, or potassium over labeled amounts," and "reasonable deficiencies of calories, sugars, total fat, saturated fat, cholesterol, or sodium"as compared to label amounts are acceptable within current good manufacturing practice.

SPECIAL EXEMPTIONS

It is reasonably clear that many small businesses simply could not comply with the labeling rules because the costs would bankrupt them. In addition, it would be desirable to avoid the mindless application of the rules to situations where publicizing the nutritional content, if any, of a semi-finished product or bulk material meeting the definition of a food would be of benefit to no one and a great financial handicap to many. A relaxation of labeling requirements in special instances was indicated, and the FDA made a few allowances for special situations. The most sweeping of these exemptions appears to be the freeing of some small businesses from certain provisions of the new labeling laws.

Exemptions for Small Businesses

In August of 1993, Congress modified the NLEA to give small food producers more time to comply with the regulations, and to fully exempt more of them. The law required most food concerns to use labels listing nutrient content by May of 1994, but it exempted companies with less than $500,000 in annual sales. Under the amended law, companies employing less than 300 workers are exempt in the first year from labeling products that sell less than 600,000 units, and in the second year from labeling anything selling less than 400,000 packages. Firms having fewer than 200 workers are exempt for the first three years from labeling items that sell fewer than 200,000 packages; companies employing less than a hundred workers are permanently exempt from labeling products that sell fewer than 100,000 units.

A business does not have to apply for this exemption, but it must be able to provide evidence it meets the requirements for exempt status if the question arises. Apparently, businesses (except those having less than $500,000 total sales or less than $50,000 in total food sales) relying on the exemption must, however, file a notice with the FDA. The declaration must state the average number of full-time employees in the twelve months preceding the time for which the exemption is claimed and the number of units sold in the U.S. during the preceding twelve months (or, if none was sold previously, the anticipated number to be sold during the time covered by the exemption).

Other Exemptions

Among the foods and food service business that do not have to comply with the NLEA regulations are:

(1) Foods served for immediate consumption, as by restaurants, delis, bakeries, and confectioneries having sit-down facilities for diners.

(2) Transportation carriers such as planes and trains, food service vendors,

lunch wagons, ice cream shops, mall cookie shops, vending machines, and side-walk food carts, all of these being business where foods are generally consumed immediately or while the consumer is walking away.

(3) Ready-to-eat foods delivered to homes (e.g., pizzas delivered to residences) and. ready-to-eat foods primarily processed and prepared on-site but not for immediate consumption on the premises and not offered for sale outside that location.

(4) Foods of no nutritional significance such as coffee beans, tea leaves, plain unsweetened coffee beverage, dehydrated vegetables, and some spices (not chili powder because it has been found to have some nutritional value!); all nutrients must be at a level that would permit a declaration of zero.

At present, raw fruit, vegetables, and fish are not required to have nutrition labeling, but labeling may be required if industry does not voluntarily comply within a time limit not yet established. Custom-processed meats from wild game and fish are not federally regulated and the NLEA does not apply.

Foods that are shipped in bulk (not in retail packages) and will be used in the manufacture of other foods, or that are to be processed, labeled, or repacked at another site do not at present come under the NLEA.

Individual food packages sold in a retail multi-unit container must bear the statement, "This unit not labeled for retail." An example would be ten packets of hot chocolate mix in an outer retail carton. The nutrition panel must appear on the outer wrapper of the package.

Certain special dietary items such as medical and infant foods, food assortments, and food supplements will be covered by other labeling regulations. Legislation has been proposed to relax the restrictions on label claims for dietary supplements, particularly to allow claims not fully supported by current scientific evidence.

OTHER PENDING PROPOSALS

According to proposals made in late 1993, the FDA intends to exempt from regulation otherwise unapproved substances if their use in packaging results in a migration to the product that would lead to dietary concentrations of 0.5 parts per billion or less. These indirect additives might include such substances as the antioxidants that have been applied to packaging materials for retarding the rancidity development occurring in shredded wheat, for example. The FDA has said certain "uses of substances in food-contact materials does not require regulation under food additive petitions" when the uses "were expected to result in very low levels of migration and that did not raise . . . safety concerns."

To qualify for exemption, the substance: (1) must not be, or contain, a carcinogen with a specified level of cancer-causing effect; (2) must not present any other health or safety concerns because its use will result in dietary concentrations of 0.5 ppb or less, or because it is already approved as a direct food additive and the proposed use is less than 1% of an FDA-determined acceptable daily intake; (3) the substance has no technical effect on the food into which it migrates; and (4) there are no adverse effects on the environment.

These limitations are such that questions can be raised as to whether there will be any benefit conferred on packagers and food manufacturers. Item (4) alone is sufficient to cause interminable legal wrangles as to its meaning and to its application in specific cases. Virtually every part of the proposal leads to a necessity for, and a certainty of, litigation for pinning down the meaning of inexact words and phrases. What is needed is not more lawyer-enriching convolutions in the regulations, but simplification and logic.

ANTICIPATED COMPLIANCE PROBLEMS

The extreme complexity of labeling regulations, the many uncertainties necessarily involved in their interpretation and enforcement, and the apparently ephemeral nature of some of the provisions, enable us to forecast many pitfalls for food producers, even those manufacturers who try their best to conform to what appears to be the existing law (perhaps especially for them). Consider the possible conflict with state laws – little thought has been given to pre-emption problems and the law does not clearly state that all contradictory state laws are to be overridden.

Consider also the unknowns involved in the analytical procedures supposed to be useful for determining the declared values of nutrients, and in the weighing and measuring of ingredients, semi-finished components, and the like, all of which affect the accuracy of nutrient content claims. What of the loss of nutrients during processing or storage, particularly if unusual events occur during any part of the manufacturing and distribution steps?

Automation of weighing, mixing, portioning, forming, and packaging may give at least some protection against unexpected delinquencies with respect to label claims. Walters (1993) explains the value of information technology in maintaining compliance with the new labeling law. He says, "Automation technology available today can ensure both batch-recipe and weight-control conformance, using off-the-shelf solutions that are sophisticated, comprehensive, and flexible enough to fit almost any food processing environment – from dry bulk-package ingredients to liquids. In addition to ensuring compliance, a system provides documentation, in the form of electronic batch records, capable of proving compliance."

BIBLIOGRAPHY

ANON. 1984. Advertising, packaging, and labeling. Office of Consumer Affairs, U.S. Dept. of Commerce, Washington, DC

ANON. 1990. Nutrient Data Base for Standard Reference, Release 9. Nat. Tech. Info. Serv., U. S. Dept. Agr., Springfield, VA

ANON. 1993. Nutritional labeling: The final regs. R-Tech Review 2, No. 1.

ANON. 1994. The NLEA and you: Options for compliance. Marprint, Inc., White Bear Lake, MN

ANTOINE, V. 1994. Spacing out. Manufacturing Systems *12*, No. 1, 6.

BOWERS, B. 1993. Small food makers seek to fatten labeling law's exemptions. Wall Street J. *91*, No. 106, B1, B2.

CAMPBELL, E. J. 1993. Labeling – 1993. Am. Soc. Bakery Engrs. *1993*, 39-43.

DEAN, B. 1993. Sorting out the new labeling regulations. Food Processing *54*, No. 8, 20-23.

DOPP, M. 1993. The compliance officer is coming. Food Processing *54*, No. 5, 18, 22-23, 26.

GIBSON, R. 1993. Label law stirs up food companies. Wall Street J. *91*, No. 106, B1, B6.

GREENBERG, E. F. 1991. Some uneven chinks in food labeling "uniformity." Packaging Digest *28*, No. 1, 26, 28.

GREENBERG, E. F. 1993A. A few observations on the first 30 days. Packaging Digest *30*, No. 3, 20.

GREENBERG, E. F. 1993B. End of the beginning for food label changes. Packaging Digest *30*, No. 1, 24-26.

GREENBERG, E. F. 1993C. FDA at threshold of new approval approach. Packaging Digest *30*, No. 13, 12, 14.

HALL, J. M. 1993A. Food labeling V. R-Tech Review 2, No. 3, 1-7.

HALL, J. M. 1993B. The rest of the story. R-Tech Review 2, No. 2, 1-10.

KOGEL, K. K., and HAYTOWITZ, D. B. 1993. Estimating nutrient values of mixed dishes from label information. Food Technol. *47*, No. 4, 69-70, 72-75.

KUSHNER, G. J. 1993. Fiber-related health claims. Food Processing *54*. No. 3, 43, 46, 48, 50, 52, 54.

MARCOE, K. K., and HAYTOWITZ, D. B. 1993. Estimating nutrient values of mixed dishes from label information. Food Technology *47*, No. 4, 69-70, 72,75.

MERMELSTEIN, N. H. 1993. A new era in food labeling. Food Technol. *47*, No. 2, 81-86, 88-92, 94-96.

ORLOSKI, A. 1994. Software satisfies Wilson's hunger for data. Packaging World *1*, No. 1, 34-36.

PEHANICH, M. 1993. New angles on nutrition and NLEA. Prepared Foods *162*, No. 3, 11.

REISS, E. S. 1977. Advertising and product development. Food Technol. *31*, No. 7, 75-77.

SHAPIRO, R. 1993. Use of databases for labeling not yet approved. Food Technol. *47*, No. 5, 32.

SMITH, J. L. (Editor) 1992. Nutrient Data Bank Directory, 8th Ed. Dept. of Nutrition and Dietetics, Alison Hall, Univ. of Delaware, Newark, DE 19715-3360.

SPERBER, R. M. 1992. Software eases label compliance. *53*, No.3, 40, 42-44.

SZMEDRA, P., and FERGUSON, W. 1991. The Delaney clause: New interpretations. Food Review *14*, No.2, 6-7.

TROIANO, J. 1994. Have you looked at our database lately? Food News for Consumers *10*, No. 4, 13.

WALTERS, R. 1993. How to comply with the new nutritional labeling law. Manufacturing Systems *11*, No. 11, 56, 58, 60.

WOODBURN, M., and RAAB, C. 1993. Product care directions on food labels: Status and needs. Food Technol. *47*, No. 2, 97-99.

FROM THE LABORATORY TO THE MARKETPLACE

INTRODUCTION

After the product (i.e., the food in its container) has been tested for consumer acceptance and shelf life, its ultimate profitability has been confirmed by standard accounting procedures, and marketing/sales managers have agreed the product can be sold in sufficient volume to yield a reasonable profit, it is necessary to set up the mechanisms necessary to produce, distribute, and sell the new item. Many new problems enter the picture at this point; only a few of them can be considered in the limited space at our disposal.

The major steps or phases preceding a food product launch were listed by Maxwell (1988) as product concept, consumer profile, market analysis, product development, consumer response, product specifications, process development, process specifications, ingredient sourcing, ingredient declaration, labeling and claims, packaging specifications, package development, package graphics design, shelf-life testing, product coding, microbiological survey, sanitation program, hazard analysis, state and local permits, federal inspections, production staffing, production scheduling, inventory control, warehousing, distribution method, marketing plan, ads and promotions, and sales program. These steps are not necessarily in the sequence of their performance, at least not for all projects. In fact, a number of the steps can be done in parallel. The list is not universally applicable, of course, but it is valuable as a rather complete set from which the steps essential to your own operation can be chosen to make a useful checklist.

Thomas (1977) described the stages leading to the introduction of new products into consumer sales outlets as:

(1) Establishing the sales potential: One of the first questions that management must resolve is whether a product meeting the original rough or temporary specifications will sell. The current market for similar items or categories must be studied.

(2) Blueprinting the product: Market research tries to discover the specific qualities, such as taste, fragrance, color, and consistency that potential customers would like to find in the product.

(3) Preliminary screening: Technical people develop many samples for examination by management. These samples are screened or evaluated by panels ranging from the CEO (only) to panels thought to have some ability to differentiate losers from winners.

(4) Final consumer pretests: After the products have been narrowed down to one or two finalists, market research tries to learn how the public thinks

the products rank against comptetitive products, i.e., similar items that are actually in, or will be in, the market.

(5) Sales testing: If a favorable verdict is received from the consumer panel, the product is put into the marketplace, usually in limited distribution, initially. During the tests, independent research firms are employed to make continuing checks of the sale of the product and of its competitors. From all this information about the consumer and the market, presumed to have been obtained under realistic marketing conditions, management can make an informed prediction of how the final mass-produced item will perform when it is introduced nationally – whether such an introduction would be worthwhile.

(6) Facilities and personnel of the Manufacturing, Distribution, and Marketing Departments are assigned to the production and sale of the approved product(s) in the quantity and at the times indicated by company strategy, the latter being based on all data accumulated up to the time the decision is made and as evaluated in relation to company goals and availablity of resources.

Previous chapters have dealt with steps (1) through (4); this chapter will cover points (5) and (6) in the above scheme, with additions as required to meet the author's viewpoints on this subject. In addition, there are several essential considerations involving mainly technical problems, such as upscaling the production processes from bench samples through pilot plant tests to full scale manufacturing, that must be dealt with by the manager of new product development.

PRESENTING THE PRODUCT TO THE CONSUMER

In the introduction of new products, there are several important elements of the selling process that are only slightly affected by product quality as perceived by the consumer but have a very important effect upon the success or failure of this phase of the new product development program. Among these elements or steps are: (1) appointing distributors for the product – if distributors are normally used in the company's merchandising efforts, (2) obtaining approval of the product by buyers for large retail chains so that it is "listed in the computer" and can be ordered by individual store managers, (3) getting initial stocking at the store level, (4) increasing consumer purchases by couponing, demonstrating, publicizing, and advertising, and (5) influencing the re-stocking rate of retailers by special promotions or other means.

All of these actions are perfectly acceptable strategems to get the consumer to try your product, but their effects must be factored into the predictions of normal consumer demand. If, for example, couponing, advertising, and promotions are conducted at levels during the test that cannot be continued after the introductory period, and sales projections are based on linear extrapolations of trends observed in the sales test, management can be misled into thinking a product can be sold at rates that are, in fact, unachievable when normal promotional efforts are in effect.

Getting Retailer Acceptance

The effectivness of brokers or distributors, if they are elements in the manufacturer's selling system, has a major influence on sales during the introductory phase of a new product. It is possible to find a broker willing to handle almost any product, but a good broker – one who will actually devote considerable time and effort to facilitating the introduction of your product, who will assist the manufacturer's personnel in presenting the product to buyers, and who will assure the timely dispersal of stocks of the new item through the retailer outlets – these distributors are in wide demand and can afford to be very choosy about accepting new products, especially those made by small companies.

Before any product can achieve significant sales through retail outlets, it must be presented to the buyers of retailer chains. Generally, a salesman or sales team from the manufacturer's staff will make the presentation – sometimes, a representative of the broker will also be present. The decisions of the buyers to list or not to list the product as an item that can be stocked in their stores will have a significant effect on the total number of outlets where consumers will have an opportunity to buy the product.

Buyers base their decisions on both personal and institutional criteria. They must decide whether a product will appeal to the consumer based on its appearance, label claims, taste, convenience, shelf life, etc. Buyers very often have no special skill in any of these fields and must make snap judgements based on their past experience with the salesman or on their own personal prejudices. If the product passes this initial scanning, they must then evaluate the item's value to the chain based on profit margins, shelf space occupied by competing items, introductory allowances, reliability of the manufacturer in making good on claims of proposed advertising and promotional activity, etc.

Most, if not all, large retailers require their suppliers to pay "slotting fees," for acceptance of a new item. These fees are upfront payments to compensate the retailer for the alleged costs of entering the product into the database that will be used by individual stores in requisitioning quantities of the item, for giving the product shelf space, and for warehousing stocks of the package. Slotting fees are heavy burdens on smaller suppliers who are probably operating on a shoestring anyway.

It appears that, in some cases, retailer chains may forego the slotting fee for a new product that strikes their fancy. Neff (1993B) says, "Though rare, some small companies get food products into supermarkets with no slotting fees at all. Others manage to negotiate reduced fees or more favorable alternatives to the up-front cash large corporations pay." He cites the examples of Glory Foods, a firm that launched a line of soul food items, and Mystic Chips, described as a unique regional product with a story to tell as well as an appeal to upscale natives and tourists alike. To generalize, it appears that a product having some unique selling point and that also provides the retailer with a large profit margin may have some opportunity for getting its slotting fees waived. Standard items with a low profit margin stand almost no chance of getting in the store without slotting fees being paid for them. Neff quotes a total cost of $3,000,000 to slot a frozen food product nationally.

A list of 43 alternatives to slotting fees has been published by the Chicago-based National Food Distributors Asssociation. Many of these involve giving the retailer free goods or advertising allowances instead of the slotting fee, i.e., they still involve up-front expenses for the manufacturer.

Preliminaries to Sales Tests

Products should not be sales-tested until a full-scale marketing scheme has been at least roughed out; it would be better to have a formal marketing plan in place at this time, but such a plan is probably almost impossible to compose in detail without having the sales test results available. It is too often the case, however, that the Marketing Department's strategies are only vaguely conceptualized at the time of the sales test, and as a result the sales test is not designed to answer some of the questions that will arise when plans for the launch are being made. Then, as Marketing starts planning for the full-scale introduction, they find that some pieces of the puzzle are not available, and they must either conduct another sales test or rely on intuition to provide the necessary guidance. Part of this problem results from the urging of top management to get a new product into the field as quickly as possible so as to maximize the company's sales, if not profits, as reported to the shareholders and board of directors as well as to investment analysts, who have such a marked effect on the price of the stock.

SALES TESTS

As previously stated, the ultimate criterion of a new product's success in the marketplace is the response of the consumer — are retail buyers willing to pay the expected price for the item, not only for their sampling purchase but for repeat purchases. Unfortunately for food manufacturers, this factor cannot be measured with a high degree of certainty, even under the best of conditions. First, the finished product in its final package must pass the evaluation of buying personnel at the retailer level — unless the product reaches the retail shelves, the consumer will have no opportunity to submit it to his final and decisive judgement. The retailer's buyers will not enter product into their stocking system unless they conclude that consumers will buy enough of the item to make it profitable for the retailer;

very often, some predetermined level of movement must be projected before a favorable decision will be made by a buyer.

The purpose of the sales test is not only to satisfy the manufacturer that the proposed new item will be profitable over the long run, but to provide evidence that will convince distributors and retailers it is to their advantage to list and stock the item.

Preliminary Considerations

A product should not be made the subject of a test market until the concept has been subjected to preliminary evaluations that take into consideration consumer preference, quality standards, stability, produceability, legal conformity, and profitability. These studies can be conducted mostly in controlled environments and will provide meaningful and reliable data on which to base an estimate of the probability of success in the marketplace.

Profitability. The question of profitability, which obviously relates to a product's contribution to the manufacturer's financial health, is a rather subjective factor since it depends to some extent on arbitrary charges to the product as decided by accounting and the CEO. It would appear that profitability is an even less objective criterion than satisfactory consumer response, but it is interrelated with the results of sales tests that may, for example, dictate changes in the product to improve its profit margin.

Of course, consumer response during market-wide selling over a period of years also depends to some extent on decisions not entirely related to product excellence – promotional allowances, advertising, couponing, sampling, etc. Ultimately, however, a new product must depend to a large extent upon its inherent excellence to generate enough sales volume to justify the continuation of its manufacture and distribution, and this level of sales must be reached within a reasonable time after the product's introduction.

Marketing mix. Some authorities use the term "marketing mix" to describe the entire complex of variables that make up a product and its marketing system. Test marketing has then been defined as the testing in the marketplace of one or more elements in the marketing mix of a product in order to determine the effectiveness of the elements (combined or singly) on a large scale. The test provides data allowing an estimation of the chances of success or failure of the product in a full-scale commercial sales campaign.

Test markets for new products are intended to evaluate all the various components of the marketing mix, but in the case of an already existing product the test market will ordinarily determine the effect of only one of the elements of the mix. The important characteristic is that market response to a product can be sampled or tried on a relatively small scale without experiencing the manifold hazards of a full-scale introduction that has been based entirely on intuition or non-selling types of consumer testing.

Kinds of test markets. According to some authorities, test markets may be divided into two broad categories: (1) Tests during which the a product and all the different elements constituting its personality and marketing approach are used, and (2) Tests of established products during which some change in an existing product or its personality or marketing approach are to be evaluated. Many of the factors involved in designing a test are the same for both types, in a few other considerations they are considerably different. Several of the important factors that must be taken into consideration when setting up a sales test of any type will be discussed in the following sections.

Planning a Test Market

A test market is a complex, costly, and time-consuming operation that requires the most careful plannning the company can manage to give to it. All too often, the sales test is rushed into operation when many of the conditions have not been fully specified.

In test markets for new products, the aim is to evaluate as many as possible of the different factors or components contributing to the achievement of a profitable level of sales. Some of the elements that will be examined are: (1) the product itself, (2) the uses to which the product is put, (3) consumer attitudes toward the product, (4) the container, including labeling, (5) the price structure throughout the distribution and sales channels, (6) methods of distribution, (7) promotional alternatives and the reaction of both retailers and consumers to them, (8) media opportunities for publicizing and advertising the product, (9) ratio of repeat purchases, and (10) how do the marketing and sales personnel of the manufacturer respond to the item initially and after the test?

Location. Since the basic justification of a test market implies that something more limited than a full-scale launch is contemplated, one of the first things to be decided when planning the test is what kind of limitations will be applied to its scope and duration. In nearly all cases it will be agreed that the geographical extent of the test will be one of the features to be limited. That is, the product will be presented to consumers in a region considerably smaller than the entire country. How will the locations of sale be chosen? Reliability and usefulness of the test results will depend on local consumer usage and attitudes toward the product group in which the new item belongs. To yield test data projectable to the whole country launch, the regions of testing must be representative of the entire country or, at least, of a very large segment of the targeted consumer group. To find one city, a metropolitan area, or a region providing this kind of cross-section is indeed difficult. It is very desirable that the area be a self-contained marketing unit, so that confusion does not arise in determining the effects of advertising, distribution, and the like on consumer response.

Small areas and large areas each have their specific advantages and disadvantages. Small towns have the advantages of requiring less investment and less effort at all levels of the company's participation, and of giving less information to competitors. Larger areas have the advantages of making it easier to analyze the effects of promotions, providing a more significant sample of the national panel, and giving a better forecast of the distribution pattern.

Some of the factors that should be considered by the market research management in selecting the location of a prospective test market are:

(1) Small- to medium-size towns may have the positive feature of representing a restricted newspaper delivery or sales area. They may not really be geographically well delimited, however, since a certain amount of spending in the surrounding area may go to the town, or to their own shops, or to other towns within easy driving distance. These towns may lack their own television stations and usually lack market research facilities. Combining several different towns in the same test may tend to compensate for these difficulties.

(2) Selecting areas based on television coverage. These tend to have adequate media resources and market research facilities, but their use will result in increased costs, and only a few such areas can be regarded as reasonably representative of the entire country.

(3) Using chains of retail stores (Wal-Mart, K-Mart, etc.). This leads to difficulties with advertising and research panels. Also, it has been said there are major problems in projecting the results to full-scale market sales, because of the presumably unrepresentative characteristics of the customer panels. However, advantages are to be found in the economics and in the reliable nature of the sales data, if taken at cash register level.

There are several other ways of delimiting test markets, as by geographic areas (counties, etc.), sales areas (in accordance to the manufacturer's sales administrative divisions), and truck delivery areas.

One list of the ten most commonly used test markets included the following areas:

(1) Erie, PA
(2) Syracuse, NY
(3) Birmingham, AL
(4) Wichita, KS
(5) Fresno, CA
(6) Spokane, WA
(7) Ft. Wayne, IN
(8) Little Rock, AR
(9) Peoria, IL
(10) Portland, ME

Promotions. The methods used for promoting the new or improved product will, of course, affect consumer response and must be considered in selecting the locations of the test. If television advertising is an element in the plan, as it will be in the majority of cases, then the area chosen will most likely be a region having restricted or dominant television coverage, or which in some other way can be isolated for this factor. Expenditure for television advertising (or number of consumers reached) can be extrapolated to national levels with little effort and a relatively small degree of uncertainty.

When the product is promoted by press advertising, circulars, billboards, mail, etc., it may become quite difficult to correlate the test results to national expectations – in terms of cost, ad quality, and number of consumers contacted.

Limitations of the manufacturing company. Every selling company will have its own special and unique limitations in staff (especially of the marketing and sales departments), financial resources, established policy (preferred ad agencies, etc.), and legal status (antitrust considerations, etc.), among others. Some, probably most, of these factors will have to be taken into account when planning the sales test, for example:

(1) Quality of the sales force. The strength of the local sales force as affected by the number, dedication, and vigor of the company's representatives assigned to that area, can affect test results profoundly.

(2) Product delivery facilities. The area chosen must be one in which the company can assure timely and efficient delivery of sample product to the assigned outlets. If the warehousing and local distribution facilities are not reliable, then out of stock conditions (or stale product) can result, with total destruction of test reliability.

(3) Company penetration of retail outlets. If the manufacturer cannot count on shelf exposure in the largest local retail stores, then sales results cannot be fully representative of that market or any market.

(4) Company's past promotions in the chosen area. Bias in results can be the consequence of heavy company promotions (or subnormal promotion levels) in the months or years preceding the test.

(5) Market bias. The company may have a history of poor performance in a given area, for reasons unknown. The unknown factors could also affect the sales test results and make it impossible to correct them when analyzing results.

Duration of the test. The length of time the test is to continue should always be decided before the first package is sold. There is a a strong tendency to terminate the test early when the results are viewed as being favorable, and an equally strong tendency to prolong the test when results seem to be unfavorable or are equivocal. If a firm decision as to the duration of the test is made before any sales data are available – this decision being based on a careful examination of previous temporal trends and sales experiences with similar products — there is much less likelihood of pressure (internal or external) to modify the test conditions.

The test must be long enough to allow sufficient time for the sell-in to stocking units, and for initial and repeat purchases by consumers. Competitive response should also be surveyed. On the other hand, overlong tests are expensive and often add little to the significance of the data. They also allow time for competitors to develop and introduce their own version of your product.

Acquiring sales test data. In former times, manufacturers tended to rely on the pattern of plant shipments of a new product as an indication of the degree of success of a new product introduction. This was not a very reliable measure, however. Shipments made during the introductory phase can be very misleading if used as indicators of steady-state sales. If the product is well-accepted by the retailers, leading to a heavy outflow as shelves and warehouses are stocked, the manufacturer may be tempted to take a very optimistic view of production requirements and start installing additional plant capacity.

This "pipeline-filling" effect has led to many costly errors in estimating future sales. When the consumer offtake is slow, the pipeline material can clog distribution channels for weeks or months, slowing plant shipments to nearly zero and resulting in the accumulation of stale product at all distribution levels. If the product has the appeal of considerable novelty, the initial sales level may be much higher (due to consumer sampling based mainly on curiosity) than any volume that can ever be achieved subsequently. If some unforeseen defect materializes, it may be necssary to take back large amounts of introductory product, leading to heavy losses which may be disastrous for a small firm. Therefore, it is very important to make a realistic estimate of the rate of consumer re-purchase and this estimate should be updated often during the test.

Another way introductory sales can be followed, especially in relation to sales of competitive products, is to obtain the reports of various marketing services that make a business of compiling reports of warehouse shipments.

More recently, the affixing of universal product codes to each package has made it possible to computerize individual sales at the checkout counter, so that actual consumer purchases can be determined on a daily basis fairly easily (provided the full cooperation of the retailer can be obtained). These data are obviously much more meaningful measurements of consumer response than either plant shipments or warehouse shipments. According to Wagner (1993), "Speed and accuracy have evolved to the point where it is possible to measure how a promotion performs across an entire supermarket chain, not just one store. Instead of a month, analysis can be done in a week."

Mistakes to avoid in test markets. A survey of marketing literature indicates the following errors are common in sales test procedures:

(1) Giving more attention to the market than could ever be given during a national launch.

(2) Overly optimistic volume forecasts, resulting in large amounts of unsalable product at the end of the test.

(3) Overly pessimistic volume forecasts, resulting in out-of-stock conditions partway through the test.

(4) Unrealistic conditions for in-store exposure and handling.

(5) Changing the criteria for failure or success after the test is concluded.

(6) Advertising projections not based on correct data.

(7) Selection of unsuitable test markets.

(8) Failure to take into consideration the nonrepresentative features of the test market.

(9) Failure to conduct adequate and pertinent consumer research.

(10) Overextending the marketing staff so that attention given to the test market is inadequate.

(11) Trying to test product and packaging variables in the same sales test.

(12) If early results make it imperative to change some characteristic of the finished product, all test parameters should be reviewed. It is hardly likely that the sales observed for the new product can be averaged with those obtained from the original product.

Some of the actions to be taken in avoiding these mistakes include:

(1) Don't allow the sales and marketing people to become overenthusiastic. The field representatives will often become hyperactive (by their individual standards) as they come under the attention of headquarter's staff and advertising agency research directors. Their level of interest and energy can seldom be translated to a national scale when the sales people in the field will be confronted with "just another" new item.

(2) Don't introduce levels of promotion that cannot be maintained in a nation-wide launch. There is always a temptation to give maximum promotional allowances and other incentives for retailers to place your test item on the shelves and then force it on consumers, as by in-store tastings, special billboarding, etc. These are also practices that lead to a volume of sales not generally reproducible in a wide-scale introduction, and certainly not in a normal year-round selling pattern.

(3) Don't use a product or package that is not truly representative of the item that can and will be offered if there is a national launch. Sales tests are evaluations of production capabilities and packaging procedures as well as marketing efforts. The item that is being bought by consumers in the sales test should be as nearly like the full-scale production model as can be achieved. Of course, it is understood that lessons taught by the sales test may need to be translated into product modifications, but an unrealistically fresh, specially produced and stringently quality-controlled, and luxuriously packaged product may give results quite different from those achievable

with an item made by the millions.

(4) Consider the seasonal effects on product sales. Holidays, hot days, cold days, vacations, and many other seasonal influences will affect the volume trends of many foods, and must be taken into consideration when interpreting the results of a sales test. Unexpected weather conditions must also be factored into your equation – floods, blizzards, and other natural catastrophes disrupt not only consumer buying but also distribution channels; off-take during such times will not be representative of sales during normal times.

(5) Always be aware that your product's sales may be inflated by its novelty appeal and your effective monopoly (in a restricted sense) during the sales test. When faced with competitive action, possibly facing an improved product a competitor has based on your initial design, sales will almost certainly be less than in the test. Competitive action during the test may be a guideline for projecting future movements by other food firms.

Going National Without a Sales Test

The easiest and quickest sales test is no sales test. It is also the cheapest way to proceed if your new product happens to be successful – the costs may reach disastrous levels if the product is a failure, however. The basic principle of the continuous roll-out or naked launch school of thinking is that a well-researched product, of a type with which the company has much experience, will give no additional significant information when subjected to a sales test. Both large and small companies are adopting this strategy according to Meyer (1994), who says they are ". . . part of a growing trend, as food companies large and small put more faith in their own research and less in the opinions of Buffalo or Peoria." It is said the same amount of work in product development, legal affairs, sales, and marketing is required whether the product goes into a sales test or is rolled out nationally. Omitting the sales test can accelerate a product by 18 months or more, according to one executive's estimate.

There are cost factors that are not comprehended in the foregoing list, and they can be substantial. Among these are the additional production facilities that will have have to be constructed to provide the much larger volume of product needed for a country-wide introduction. If the product is a failure, the quantities of unsalable packages that have to be disposed at much reduced prices or even at net losses may reach unmanageable levels. The effect of a failure on the attitudes of the manufacturer's brokers, wholesalers, retailers will be much milder if they were told at the start that the new product is in a test.

Marketing and sales heads, as well as CEOs, are anxious to get increased volume, however, and many of them are quite willing to gamble the company's funds in attempts to rapidly increase sales.

LESS COMMON WAYS OF PROFITING FROM NEW IDEAS

The small operator, whether an individual entrepreneur or the co-owner of a small food company, often encounters seemingly insurmountable problems in getting financing for the development and introduction of a new product. The physical realization of a new product concept becomes prohibitive not only in terms of available funds but also in time. In the following section, two alternative ways of reducing the need for capital investment and for decreasing the time needed to get the product into the marketplace will be explored.

Selling Concepts to Functioning Companies

The inventor or developer of a new item or process may decide that no reasonable opportunity exists for setting up manufacturing and merchandising facilities for the new product under his or her ownership or direction. The innovator may suspect that adequate financing for such an operation will never be forthcoming under any conceivable scenario, or perhaps the inventor does not regard himself as being enough of a manager to carry on the day-to-day operations of a new business. Among the options available for gaining some financial return for the new product is selling the idea to a company making similar products. For example, RJR Nabisco management might be approached with an idea for a cracker-sandwich having mushroom-flavored or caviar-flavored filling, Coors might be receptive to the idea of marketing your pretzel-flavored beer, etc.

It is probably only rarely that top management of a large consumer products company will consider ideas presented by individuals from outside the company. In fact, they all too rarely give any serious consideration to ideas from inside the company. Many of the submittals to the CEO or his assistant will immediately be bucked down to lower levels within the company, probably to someone in R&D who is having trouble getting his or her own ideas put into the system. Nonetheless, there have been a few cases where unaffiliated innovators have had their ideas picke up by large companies, It is worth a try, if all else fails.

Specialized Distribution Niches

If the entrepreneur is willing to begin his business by selling the new product on a very small scale, there are several specialized outlets for which personal attention is an important factor and which are therefore particularly suitable for new product introductions. Of course, we have all heard the legends of housewives who began selling their home-baked bread, cakes, pizzas, cookies, or pies house to house and then developed the business into multi-million dollar enterprises (Pepperidge Farm, Brownberry Ovens, Marie Callender, Rose Totino, Mrs. Smith's Pies, Mrs. Field's Cookies, etc.).

Whether it is possible in our present regulatory and competitive climate to duplicate such scenarios, or, indeed, whether the companies even evolved as has been described, is debatable – though the stories do have inspirational value.

To take advantage of most of the opportunities that will be listed here, it will be very helpful if the production facility is close to a metropolitan area. Otherwise, expansion of the merchandising effort beyond a very small scale is virtually impossible using these methods.

Distribution from door to door, or from office to office, is still a method by which products made in the home kitchen enter the commercial realm. In most localities, laws and regulations make this type of food distribution difficult, perhaps costly, and always risky. It is best to avoid it, except as a last resort or as a test on a very small scale of consumer reception. It should also be mentioned that this type of selling is not adapted to the personalities and plans of most entrepreneurs, who like to work on a larger scale.

Many delicatessens, neighborhood bakeries, specialty or boutique restaurants, meat departments, produce sections, and caterers are receptive to the idea of buying unique or gourmet foods from small suppliers. In some of these instances, packaging and promotional complications are essentially non-existent. Dollar volume per hour of selling effort will be small indeed, but the experience obtained will be of value not only in perfecting the product but (if the product is well-received by consumers) as a selling point when introducing the product to supermarket buyers. It is best to look upon this type of operation as a sales test rather than as an ongoing business.

As a means of increasing their product line, specialty food distributors will often consider cooperating with smaller manufacturers. Uniform packaging with legal labels, uniform and shelf-stable products, and reliable stocking practices will be needed when supplying these distributors.

If your product is suitable for use in any kind of foodservice operation (restaurants, hospitals, university dining halls, prison feeding kitchens, airline commissaries, etc.), thought should be given to approaching the directors of these services to determine their willingness to purchase the item. Headquarters of large chains of restaurants provide a potentially large market for privately made food products, but often take a long time to evaluate the item, and then may decide to make a similar product in their own captive production facilities.

Mail Order Sales

The use of mail order sales as an initial outlet for new products designed and manufactured by the small independent company seems at first glance to have many advantages. The entrepreneur will see that the need for expensive retail outlets is avoided and/or costly introductory presentations and allowances for supermarket stocking rights are eliminated. In general, personnel problems are also reduced, and roll-outs can be controlled and monitored easier. It is even possible that certain costly and delay-causing regulations can be circumvented. There are, however, many problems with this merchandising alternative. Among them are:

1. Storage stability. Some products are not adaptable to this method of merchandising because the food needs to be held under special conditions of temperature and humidity in order to maintain satisfactory quality, or it may be peculiarly susceptible to mechanical damage; in some cases, the product deteriorates fairly rapidly in spite of optimum storage and handling treatment. The distribution conditions encountered during delivery through mail, UPS, or even Federal Express services cannot be depended upon to give uniformly reliable handling of delicate foods. Obviously, ice cream can't be satisfactorily mass-merchandised by a mail-order operation, and it is also true though not so obvious that a large proportion of of foods, including many baked foods and snacks, are not sufficiently resistant to microbiological spoilage, staling by other means, and mechanical damage to allow successful distribution via the postal service, UPS, etc., unless special (and often very costly) protective measures are used, such as baking in cans or freezing the product solid, packaging it in insulated containers, and distributing to the consumer via overnight delivery services,

2. Advertising, mail solicitations, publicity campaigns, and other efforts aimed at securing orders can be very expensive. Designing and printing the catalog and/or other mailing piece gives rise to up-front costs that are often prohibitive. The catalog may need frequent re-designing as sales results accumulate, adding greatly to the cost. Purchases of mailing lists and advertising space, postage, and other essential expenses are continuous and heavy drains on funds of the mail merchandiser.

3. Often, the new product is not enticing enough to motivate potential buyers to go through the added effort and the waiting time involved in mail orders or even in phone ordering. These inconvenience factors must be overcome by a product or delivery feature or other characteristic of strong appeal to a limited but relatively affluent group. Possibilities include dietary foods meeting the needs of small percentages of the population who find their choices in supermarkets are very limited or even non-existent.

Examples that might be considered are gluten-free bread (perhaps canned), truly sugar-free cookies, certified kosher snacks and baked products, vegetarian versions of sausage rolls and steak and kidney pies, and high-fiber dessert items. It is recognized that all of these materials are currently available in specialized markets or in certain locales, but in some markets mail order merchandising would provide the consumer the advantages of wider choice and higher quality. Reaching these markets would require the purchase and use of specially sorted mailing lists and narrowly targeted adverising campaigns.

Foreign foods having recognizable quality differences and actually made in the country or region of origin, such as kolachy made in Poland, matzoh ball soup made in Israel, haggis made in Scotland, cous cous from

Morocco, and Danish pastry made in Denmark (yes, I know that the original Danish pastry wasn't made in Denmark and that the present idea of the product doesn't come close to the original) might be well received in wider markets than the ethnic groups that make up the obvious targets. Of course, merchandising of this type is being done already; imagination will be required to develop or identify new items in the field.

Specialized delivery services such as the mailing of gifts directly to recipients during holidays or on birthdays, on wedding anniversaries and other anniversaries, at the birth of a child, at quincenaras, bar mitzvahs, and christenings, possibly with gift wrappings and always with cards identifying the giver, is a device currently being worked almost to exhaustion, but some of the successful marketers now operating in this field might be willing to take on your specialized new product, giving you a relatively quick test market. Don't expect to make much profit in this wholesaling position, however, and do expect your idea to be stolen in the next season if it has been a good seller. A pattern to follow is Swiss Colony, a very astute marketer of high quality foods at prices that do not seem particularly outrageous.

4. Setting up a plant to produce the new food item will be expensive and time-consuming no matter what method of market introduction is used. Purchasing specialized equipment or very costly equipment of any sort should be avoided since the need may be obviated by a test failure or by a change in product characteristics leading to changes in production methods that rule out use of the kind of machinery you have bought. Using manual labor to perform all or most of the manufacturing steps should be the choice in the initial scale-up – don't even think about automation at this stage.

Small operators should consider having the new product made by contract manufacturers. If the processing and packaging methods needed for making the new product can be readily adapted to production lines already in operation at other companies, contract manufacturing may greatly reduce the capital investment needed to make the initial stock of a new product. Furthermore, the costs of training personnel, putting the factory into compliance with regulations, identifying sources for ingredients and other raw materials, and setting up a distribution system can often be greatly facilitated by making use of a company already producing food products of the same general type as the new item. Of course, confidentiality is lost, and you are likely to find that your supplier becomes a competitor if you move your production elsewhere.

5. In a number of respects, consumer relations become more difficult with mail selling. Complaints about product quality and timeliness of delivery (including allegedly lost goods) will prove to be greater than for over-the-counter selling, and often there is no remedy for the problem except a refund. A relatively high level of disputed billings and other collection problems can also be expected. Charges to credit cards always involve some delay, lost or erroneous billings, extra work, and discounts.

SCALING UP PRODUCTION

Serious problems with product quality can occur at any stage in the usual development sequence of bench test samples, pilot plant production, and full-scale processing that leads from the first set of samples made in the laboratory with careful attention to detail and much hand manipulation to the automated production line putting out hundreds or thousands of packages a minute. A process that seems very simple and easy to control within narow ranges in the laboratory may prove to be difficult if not impossible to duplicate in the pilot plant, much less in the factory. It is important to learn of these problems before sales tests occur. Ideally, the sales test samples should be run on exactly the same type of equipment that will ultimately be used to make product for the national launch, if such occurs. Furthermore, the processing line should be running at about the same speed it is expected to achieve in full-scale production, at least for part of the time, so the range of product quality that will be experienced in mass production can be properly assessed.

For many small companies, however, there is no prospect of up-scaling the processing operation beyond installing a few pieces of simple equipment to mechanize a few stages of what has started out as an entirely manual operation. Often, the first operation to achieve any degree of automatic processing is the packaging step since the general appearance and thus the consumer's first reaction to the product is much improved by continuous packaging equipment.

To obtain at least some experience with line processing, it may be desirable to seek out the help of organizations that maintain sets of equipment for pilot plant testing on a rental basis – some of these suppliers will have lines that approach full-scale manufacturing conditions. Among the better known test plants is that maintained at the American Institute of Baking at Manhattan, KS; the Department of Milling Science and Technology at Kansas State University, same city, may have some equipment of the same type though arranging to use it may be more difficult than dealing with the AIB. The latter organization has small scale production lines for most baking operations including a cookie line. They also have a staff of technologists that assist in running the tests and may be able to suggest improvements in your process or formula.

Other locations for specialized pilot plants include, at various locations in Grand Rapids, MI –

Baker Perkins Food Extrusion Systems for flaked breakfast cereals, co-extruded snacks, baby food, pet food, or licorice confectionery.

Werner Lehara Bakery and Snack Technology Center for cooking, blending, extruding, and many kinds of cookie processing.

Rose Forgrove for packaging machines and equipment.

Many equipment vendors will have functioning full-size units available for testing by potential buyers, either at their main plants or in the plants of some of their customers, but these arrangements seldom give useful experience in the complete sequence of processing steps and often operate under restrictions making them of little value except, perhaps, as a comparison with competitive units performing the same function. Transporting partially finished products to the next set of equipment, which may be half-way across the country, leads to many uncertainties. Confidentiality may also be a problem.

BIBLIOGRAPHY

ANON. 1987. Technical centers expand product development capabilities. Food Technology *41*, No. 3, 110-111.

BEST, D. 1993. The three legs of commerialization. Prepared Foods *162*, No. 11, 73-74.

CLARK, P. J. 1987. Food manufacturing: Opportunities for improvement. Food Technology *41*, No. 12, 56-58.

CLAYTON, J. T. 1987. Flexible manufacturing systems for the food industry. Food Technology *41*, No. 12, 66-68, 70.

DREIER, W. 1988. Avoiding the common problems of process scale-up. Prepared Foods *157*, No. 6, 131-132.

GALANTE, S. P. 1987. Merger of two bakers teaches distributors a costly lesson. Wall Street J. *Sep. 14, 1987*, p. 23.

GEREN, G. S. 1984. Proprietary rights in start-up companies. Research & Development *12*, No. 12, 29

GROSSMAN, L. M. 1993. Coca-Cola plans to shorten time of soda launches. Wall Street J. *92*, No. 99, C-10.

KATZ, F. 1988. In test marketing. Cereal Foods World *d33*, No. 3, 314.

KIRK, J. R. 1988. Easing the way from laboratory to marketplace. Food Technol. *42*, No. 2, 68, 70.

LAHVIC, R. 1988. From UPC to DEX/UCS/DSD. Bakery *23*, No. 3, 15.

MANS, J. 1987. Pilot plants bridge new products and production. Prepared Foods *156*, No. 8, 175, 178, 180.

MAXWELL, D. 1988. Small processors benefit from integrated planning. Prepared Foods *157*, No. 11, 92-93.

MEYER, A. 1994. Food companies graduate from test marketing. Food Business *7* , No.1, 30.

MULLICH, J. R&D vs. marketing: Time for a truce. Food Business *6*, No. 13, 9-12.

NEFF, J. 1993A. Line extensions: Enough already? Food Business *6*, No. 13, 20-21, 24.

NEFF, J. 1993B. Sometimes mom & pops dance without paying the piper. *6*, No. 14, 14.

SALANT, A. S. 1987. A cooperative manufacturing technology program for the food industry. Food Technology *41*, No. 12, 71-72.

SALVAGE, B. 1987. How to introduce a new product to foodservice. Prepared Foods *156*, No. 8, 167-169.

THOMAS, R. H. 1977. Packaging concept development: A marketing/engineering interface. Package Develop. & Systems 7, No. 4, 22-25.

WAGNER, J. 1993. The scanner revolution: Food marketing by the numbers. Food Business *6*, No. 13, 27, 29.

WILLIAMSON, T. H. 1987. Super warehouse markets: what they are, how they operate, why they succeed. Cereal Foods World *32*, No. 6, 420-421.

chapter twelve

USING COMPUTERS IN PRODUCT DEVELOPMENT

INTRODUCTION

It is not possible to cover in the space available for this chapter all the important features of computer design and computer applications relevant to the various procedures involved in new product development. The rapid advances being achieved in computer hardware and software would make much of the discussion seem out of date soon after the book was published, if their current status and specifications were reported in detail. Therefore, stringent limits had to be set on the material presented. The goals that have been adopted are to give a basic introduction to the subjects under consideration, describe a few actual applications, and suggest sources of additional information for those readers who want to learn more about topics of particular interest to them.

In following through on this plan, brief summaries of the principles of computer operation will be followed by moderately detailed discussions of the forms of computerization that have been found most useful for laboratories, pilot plants, and production equipment and, thus, most likely to be encountered during a product development project. Some of the positive and negative factors connected with computer installations are described.

Articles listed in the bibliography will supply interested readers with reviews of topics treated much more superficially here.

COMPONENTS OF COMPUTER SYSTEMS

All of the components of computer systems (or, virtually all) can be divided into the categories of hardware and software. In most cases, the applicable classification is readily discernible, the hardware being all of the machinery and devices while the software is everything else. More particularly, software is the arrangement, design, format, or pattern of electrical, electronic, and magnetic signals directing the performance of some meaningful operation by the hardware, these operations being represented at their most basic level by an elaborate sequence of on-off (i.e., digital) switching.

The Hardware

Essential elements. The essential elements in any normally-functioning computer hardware system can be categorized as input devices, memories, processors, control units, and output devices. Brief descriptions of these units are given below.

(1) Input devices. These devices are needed to transform some aspect of an event happening outside the computer into a set of electrical impulses that can be processed by the computer. The "event" may be the pressing of a button, fluctuations in the intensity of a beam of light, alterations in dimensions of a strain gauge, or any other phenomenon that can be translated into an electrical signal. A keyboard is an input device.

(2) Memories. Information storage facilities are made in several different designs. The most common formats can be described as semiconductor arrays on a chip or magnetic domains on a disc or tape. Recently, there has been a great increase in the use of optical discs that store data in the form of physical changes on a plastic or metal disc allowing for sensing by optical devices. The memory of whatever type must be accompanied by a set of instructions, an "operating system," that tells the computer how to perform its functions.

(3) Processor. A collection of microscopic electrical circuits on silicon chips (plus associated connections) that processes the incoming signals to yield sorted and identified outgoing signals. Often described as the "brain" of the computer.

(4) Control unit. Governs the supply of data to the processor and to the output devices. Also controls the input of program instructions.

(5) Output device. Translates computer code into a display on a monitor, printed characters on a sheet of paper, or a control signal that affects some machinery function or performs some other useful action.

Auxiliary equipment. There are numerous pieces of auxiliary equipment, either necessary or optional, that aid in networking, restrict access to confidential information, allow use of phone lines for transmitting and receiving coded data, protect the system against environmental hazards (such as static electricity), and guard against excessive fluctuations in the line current.

Many personal computers (PCs) are sold as a more-or-less complete system that includes the keyboard, internal memory, one or more flex disk drives, and a hard disc. Nearly all of these packages will also include a TV-type monitor, and many of them will include a printer. as well as the cables that connect these output units.

There will be an internal permanent memory that includes the disc operating system (DOS); this establishes the basic functions needed to access programs. All added software must work in conjunction with the DOS, although the latter can be modified out of all recognition by software additions that take control once the operation has been set up.

Personal computers of the most advanced designs can have a place in a plant's data analysis and process control functions, although their memory, speed, and networking capabilities may not be adequate to handle all the tasks encountered in a large or particularly complex plant.

When investigating larger (than PC) computer installations, one will encounter the terms micro-computer, mini-computer, main-frame, and supercomputer. The dividing line between these classes is not clear, but the progression in the series as stated is from relatively small to very large.

Generally, the larger computers are not available as packaged systems; they are assembled from separate units each of which must be obtained from a specialized supplier. PCs have rapidly increased in power and versatility to the point where they now do tasks formerly thought to require mini-computers or even mainframes. A higher level computer should never be considered if the work can be performed by a PC or an array or network of PCs.

Installation of minicomputers or mainframes always requires the participation of a team of computer experts – persons having the requisite skills are seldom available in-house. A consultant firm will often be retained to design the system, purchase the separate machines, assemble them into a functioning computer and controller, install the programs, and train plant personnel in use of the system. This process can be very expensive, disruptive to plant operations, and disappointing so far as results are concerned. It is absolutely imperative to establish rigid cost, time, and results parameters when hiring a consultant firm on a turnkey contract, and to verify the consultant's past achievements and present abilities. Furthermore, one or more computer-knowledgeable and reliable persons should be assigned to constantly police the consultant's activities and results and to report frequently to top management on progress and prospects.

The Operating System

A completely unprogrammed computer is incapable of recognizing keystrokes on its keyboard or displaying messages on the monitor screen. Most computers are designed so that, when first turned on, they automatically begin running a program found in read-only memory (ROM). This program in turn enables the computer to load a larger set of information contained in ROM or on a hard disk, floppy disk, or tape. These more extensive instructions (called an "operating system") direct the performance of the repetitive and basic operations necessary for proper functioning of various kinds of add-on software that make the whole computer complex of some practical value to a commercial user.

The operating system enables the computer to recognize and follow commands typed by the user. In addition, it provides built-in routines that allow a user's programs to perform input-output operations without specifying the exact hardware configuration of the computer.

In the real world, an operating system will bear a name such as MS-DOS, UNIX, or OS/2, with a designation of its version, such as MS-DOS 3.1. These are standardized platforms of which millions of copies have been sold.

A computer running under one operating system cannot, usually, run programs designed for use under another operating system. Individualized operating systems can be developed by consultant firms for highly complex or highly specialized operations; they may cost the client hundreds of thousands of dollars to develop and may never work to the satisfaction of their ultimate user, resembling in the latter characteristic the mass-merchandised operating systems.

The vast majority of IBM-compatible PCs in operation today have been purchased with some version of MS-DOS already installed. These Microsoft products have always been recognized as being replete with problems and limitations, but the phenomenally successful marketing tactics of the company have enabled it to gloss over or obscure these defects. Many more applications programs, as for accounting, desktop publishing, graphics, and database handling, have been written for MS-DOS than for any other operating system. Non-IBM computers (in the desktop world, this will mainly include Apple computers) require different operating systems and, of course, different applications software.

UNIX is an operating system, or family of operating systems, the basis for which was developed at Bell Laboratories. According to one review, the main features of these systems include:

(1) Highly modular, structural design. The entire system is defined in terms of a relatively small number of sub-routines; UNIX can be implemented on any computer by carrying out these routines. The operation of the whole system is based on a relatively small number of principles carried through very consistently; there is little arbitrary detail for the user to learn.

(2) Extensibility. It is not unusual for an operating system to provide the user with the ability to create special commands of his own that execute sequences of ordinary commands stored in files, but UNIX provides a full-fledged programming language for this purpose (the "shell") containing IF, WHILE, and CASE statements similar to those of Pascal, as well as extensive facilities for string handling.

(3) Input-output redirection. Any program invoked by a command can be told to take its input from, and direct its output to, a file other than the user's terminal.

(4) Tree-structured directories — almost essential if a single user is to keep track of more than a few dozen files, and most desktop computer users (at least in business or scientific environments) need more files than this just for their applications programs.

(5) Many UNIX commands called filters read a file and output a copy of it that has undergone some simple transformation, such as removing repeated lines. Very complex operations can be accomplished by stringing together filters through pipelines.

LABORATORY INFORMATION SYSTEMS

Much has been written about the use of computers in laboratory information management systems (LIMS). It is generally conceded that every laboratory needs, as a minimum, some sort of personal computer to acquire data, massage it, and deliver it in statistically-analyzed (and preferably graphically enhanced) form. There are available a number of commercial off-the-shelf systems, which are said (often inaccurately) to be installable by the purchaser as turnkey LIMS operations. A good basic system of this type is a few orders of magnitude less expensive than a specially engineered setup and can be put into operation with much less delay, but it may not fill all the needs of a plant, especially if automated equipment is to form part of a final network.

Add-on boards are offered by some companies to expand the capabilities of PC/XT/AT type computers destined for laboratory work. One supplier advertises that their boards will "turn your PC's into powerful multifunction analysis instruments, supporting a variety of industrial process control and data acquisition applications." One version is said to give 16 channels of 12-bit analog-to-digital conversion, 2 channels of 12-bit digital-to-analog conversion, 24 bits of TTL-compatible I/O, 5 timer/counters with 1 microsec resolution, menu-driven software, and support for all MS-DOS languages.

The successful utilization of a computer often depends as much on personnel attitudes as it does on the characteristics of the equipment. Misroch (1988) described a strategy or plan for implementing the installation of an LIMS. The salient requirements of this strategy include:

(1) A detailed functional specification.

(2) The assignment of a system manager and/or administrator. This is mainly necessary when multi-user systems or a PC network is involved.

(3) A detailed system implementation plan. The LIMS will alter some of the laboratory's operations, provide new capabilities, and give new and different responsibilities to staff members.

(4) A suitable growth projection plan.

(5) All networking and application interfacing requirements. This includes direct instrument interfacing with the LIMS, interfacing of robotics applications, and interfacing to manufacturing and accounting systems.

(6) A sample flow diagram indicates how samples enter the laboratory, how work is scheduled, how results are entered and validated, and how reports are formatted and distributed.

(7) Information flow diagrams indicating what information is required for each specific purpose, such as sample identification, reporting, and retrieval.

There is no doubt that many laboratories are successfully using personal computers in an LIMS, but larger computers may be justifiable in certain situations. The number of software programs available for LIMS application is large and changes constantly. It is good practice for the lab manager who is designing the system to survey as many software packages as time permits. The efficiency and success of the whole system depends on the selection of software applicable to problems that will be encountered in the specific laboratory. And, most importantly, lab personnel must be able to comprehend the software and to put it into operation with a practical amount of training and experience.

The full advantage of data processing by LIMS will not be achieved until the computer is interfaced with most or all laboratory instruments. In order to interface an instrument to the computerized network, several problems must be solved; a list compiled by Davis et al. (1988) includes these factors:

(1) Compatibility. The instrument interface must be compatible with the LIMS computer. For example, an instrument having a parallel interface can't be connected directly to a serial port on the computer. Instrument and computer must be configured in exactly the same way.

(2) Synchronization. The computer must be synchronized with the instrument output so it is always ready to receive data from the instrument. Synchronization may be achieved by defining commands to be sent to the instrument and adding markers to define the start and end of messages or packets of data returned by the instrument.

(3) Recognition. It must be possible for the LIMS to recognize data it requires from the input device. The ability of the computer to recognize this information is dependent on the type of report the instrument is able to supply. The system must be programmed with software enabling it to read a report, e.g., to recognize key words that define the location of a data item in the report.

(4) Identification. After it recognizes the data acquired from an instrument, the LIMS must be able to identify the data with appropriate records in its database. The pieces of information required to identify a data item or result are (a) the name or identifier of the sample, (b) the name of the test or analysis, and (c) the name of each piece of data or result if more than one is produced during an analysis.

(5) Implementation. A method of operation must be chosen to define how the instrument fits into the LIMS environment.

It is important for all those concerned with establishing a LIMS to understand that constant changes are occurring in the science and technology of computer design, and that a system procured today may be outmoded tomorrow. Cooper et al. (1992) pointed out, "The life of a LIMS is finite and that needs to be communicated to both the users and management. Our task is to maximize that life to insure adequate payback. Current systems are built with more adaptability, due to modular design, but they will not last forever either."

THE AUTOMATED LABORATORY

Computer science has evolved to the point where the great majority of laboratory bench operations could be automated, although the cost of equipment seems prohibitive for many applications. Because automation reduces the need for skilled labor, often provides greater safety to workers, and can perhaps increase the accuracy (or at least the uniformity) of analyses, there is an increasing interest in automating every aspect of laboratory work that is capable of being handled in this way. New product development, as well as QA/QC, can benefit from the efficiencies and increased accuracy resulting from automation.

Good Automated Laboratory Practices (GALPs) have been issued as a draft document by the EPA. It is assumed that eventually a set of reviewed and approved recommendations will become available. The GALPs define appropriate organizational structures, testing procedures, security controls, and documentation requirements for an automated laboratory. They are intended to provide a starting point for evaluating the efficiency and data reliability in testing facilities. Many federal agencies will doubtless incorporate the interim or final guidelines into the specifications to be used by suppliers of goods and services to the government. So far as new product development for commercial goods is concerned, GALPs may be considered as interesting suggestions for labs engaged in this sort of work.

Fowler (1988) said there are between 1,000 and 1,500 robots in the 75,000 chemical and physical testing laboratories in the U.S. Probably less than 100 of these robots have built-in capabilities for functioning in a coordinated way with an instrument controller or LIMS. In the six years since Fowler wrote his article, the number of robots must have increased several times.

Fowler says there are three aspects of laboratory automation (information – data and methods – flow, instrument control, and robotic sample handling) that can be integrated both conceptually and practically. Four key strategic methods that can be taken to implement integrated automation are: (1) Adopt a broadly accepted comunications standard; (2) Use instruments with built-in intelligence and communications capability; (3) Use a broadly accepted computing platform; and (4) Use modular software split at the highest level by function.

COMPUTERS IN PRODUCT DEVELOPMENT

Bibliographic Searches

New product development of any significant degree of sophistication will require an extensive study of the literature to find out the state of the art and to unearth any prior developments of the specific type of item under consideration. These searches can take an enormous amount of time and may, in the event, fail to discover the most relevant publications. Computers excel in this type of search. Several compilations of data important to the food product developer have been listed elsewhere in this book, and can be located by using the index. A few more that should be considered are discussed in the following paragraphs. and can be classified in at least two different ways: (1) using a CD reader to search discs containing compilations of relevant data, and (2) using a computer modem to search the large and diverse resources of various commercial database suppliers.

Data on discs. A computer with a CD-ROM drive and suitable software will give the computer operator access to large collections of data available on optical disks; examples are the Oxford Unabridged Dictionary, indexes of many scientific journals extending over decades of publication, and encyclopedias of various kinds. Disks of interest can be purchased from government agencies as well as commercial sources.

The FDA Office of Information Resources supplies "FDA on disc," said to be an easy to use compilation of manuals and documents. Documents currently included are compliance policy guide, the FD&C Act and related laws, new regulations, talks and papers by FDA personnel, press releases, inspection operations manual, and FDA phone book. Other material will be added. A subscription for four quarterly disks, quick-search software, and user's guide, is $300.

World Resources Data Base Diskette provides data on global environmental conditions and trends involving energy, forests, water resources, population, food and agriculture, atmosphere, and other topics.

Environmental Chemicals Data and Information Network is a CD-ROM that contains data on the ecological impact, regulatory requirements, and potential environmental and workplace hazards of over 7,500 substances.

SigmaStat is a statistical package that includes *t* tests, analysis of variance, correlation, rates and proportions, nonparametric methods, regression, power, and sample size.

CAChe is primarily for chemists who are involved in designing new compounds; it provides them with insight into possible reaction mechanisms, etc. Call 1-800-547-8949.

The Federal Register is offered on CDs by Counterpoint Publishing Co., Wellesley, MA

Genesis nutritional labeling and formulation software can be obtained from ESHA Research, P. O. Box 13028, Salem, OR 97309.

Accessing on-line databases. The addition of an appropriate modem will allow a computer to be connected to a telephone line so as to gain access to data banks that will (for a fee) provide vast amounts of information on scientific, economic, engineering topics, and other areas of interest. This information will generally be more up-to-date than data on CD-ROMs.

Software of the proper kind is needed to take advantage of the many data banks that can be accessed by modems. A commercial program, called Scimate by its developer, eliminates the need to know the individual command language of each data base supplier. A search can be typed on an appropriate screen and then the search is sent electronically to the data base computer. As the search is completed, the program allows the user to save only selected references, store references in subject or project files, print bibliographies, and report and use the references in word processing programs.

An example of the databases accessible by food scientists is Chemical Journals Online (CJO), a subsidiary of the American Chemical Society. This offers a full-text computer-powered on-line search system. A CJO User Manual as well as an occasional CJO Bulletin are provided. This system includes not only all the journals published by the American Chemical Society, but many other chemical and chemical-related journals, truly an enormous resource for anyone working in these fields.

International Food Information Service (IFIS) provides online and CD-ROM versions of Food Science and Technology Abstracts. They state that about 400,000 entries from 2,000 sources (1969 to date) are now accessible. For specific information on packaging, a special database, PSTA, is available. Call or write National Food Lab, 6363 Clark ave., Dublin, CA 94568.

Other databases and software programs that might be of interest to food product developers include:

(1) DIALOG is reputed to be the world's largest databank; it is operated by Dialog Information Services, Palo Alto, CA. Among the 425 or more databases available on this system is the store of over five million world patents.

(2) GEnie, operated by GE Information Services (1-800-638-9636), offers a wide range of services, including electronic mail, online shopping, educational and recreational programs. It also gives access to many computer shareware programs (as do some of the other services mentioned here).

(3) Prodigy is a subscriber service (jointly developed by IBM and Sears) providing a wide variety of functions, most of which are only peripherally related to food product development.

(4) CompuServe, owned by H&R Block is also a subscription service, it is strongly oriented to computer experts, but has many bases of more general interest. Call 1-800-848-8199.

(5) America Online is somewhat similar to Prodigy in its operational details, but it has a different array of services.

(6) STN International would be of greatest interest to chemists because of its apparent concentration on structure and properties of chemical compounds. Phone 1-614-447-3600 for more details.

The Federal government sponsors several electronic bulletin boards, among them:

(1) National Agricultural Library, Agricultural Library Forum (ALF), which concentrates on agricultural research information, and can be reached at 1-301-504-5113.

(2) Bureau of the Census, Bureau of Economic Analysis provides business and industry information, economic analyses, and agricultural reports. Contact at 1-301-763-1580.

Computer-aided Experimental Design

As pointed out by Floros and Chinnan (1988), "The formulation of a new product or the improvement of an old one, the development of a new process or the optimization of an existing one, all can be better understood using appropriate statistical, mathematical, and computer techniques." They described the use of computer graphics in obtaining optimum solutions. According to these writers, the four most important elements that need consideration before seeking optimum solutions are:

1. Defining system parameters. Factors affecting the behavior of the system (such as time and temperature) in question have to be identified.
2. Describing the system mathematically. An equation must be developed to describe the relationships between factors and results.
3, Collecting data. An experimental design must be developed and carried out.
4. Fitting data to a model. After the experiment has been performed and adequate data collected, the approximating function is fitted to the results and coefficients determined.

Floros and Chinnan describe in more detail than can be reproduced here a possible graphical method for seeking optimum solutions and provide an example of how the method can be applied to seeking the best conditions for chemical peeling of tomatoes.

In an interesting recent paper, Joglekar and May (1990) describe how a frozen pizza might be developed through a series of computer-designed experiments. Their comments are applicable to many other kinds of projects. In general, the question of accelerating product and process development requires deciding which knowledge is relevant and how to obtain that knowledge rapidly. They suggest that four steps are necessary to accelerate development. First, use a proper overall framework to address the product and process development problems in order to avoid major recycles in the development process. Secondly, recognize what constitutes relevant knowledge and use an appropriate strategy of experimentation in order to generated that knowledge rapidly; they estimate that use of well-planned experiments can reduce the testing effort by a factor of two or more. Thirdly, focus technical knowledge to achieve simplification, thus accelerating the development effort. The fourth step is to use computer-aided approaches for planning and analyzing experiments; this will allow developers to concentrate on critical technical questions while delegating the statistical aspects of experiment design and analysis to a computer.

Collecting and Analyzing Sensory Test Data

Taste tests and other organoleptic tests are valuable tools for predicting consumer acceptability of new products. They are, however, time-consuming and involve many areas of uncertainty. Some of the problems can be overcome or at least ameliorated by use of computer programs designed to plan the tests, analyze the data, and prepare reports.

A computerized sensory analysis procedure based on an IBM-PC compatible local area network was developed by Findlay et al. (1986). Panelist input was simplified by use of a light pen and an interactive questionnaire program. The system was integrated to allow preparation of descriptive, hedonic, triangle, structured, and unstructured ballots; registration of panelists; collection of data; and statistical analysis and report generation. Among the benefits achieved by the system are the simplicity of response for panelists, flexibility for the sensory analyst to design questionnaires, and elimination of the time-consuming manual scoring and data manipulation involved in conventional sensory analysis.

In one use of the system, 26 panelists were divided into four separate groups to evaluate nine sugar cookie formulas. The purpose of the test was to determine optimal dextrose and sucrose levels. It required less than 10 minutes to evaluate each of the three attributes presented to the panelists. Four samples were presented each time. The accumulated data were analyzed and reported in less than 5 minutes.

To illustrate integrative capabilities of the system, the data set was transferred to a mainframe computer for multiple regression analysis. The resultant data were used to generate three-dimensional response surface plots indicating the effects of the two independent variables on each attribute and permitting formula optimization.

Graphical and Statistical Analysis

In the course of most, if not all, product development projects, collections of data are statistically analyzed and charts prepared in order to make some sense of what otherwise seems to be a chaotic mass of detail. Upper management executives seldom have the time or the attention span necessary for wading through all the material the technical people have accumulated since the last meeting. One bar graph and one pie chart just about exhausts their patience. Preparation of tables and charts from statistically-analyzed data is one of the things computers do best. A selected and tested version of standard accounting software, perhaps slightly modified, can convert large amounts of data into forms easily comprehended by those readers less technically inclined.

Many software programs adaptable to PC use can produce simple charts and graphs that give interesting and, perhaps, even informative representations of data collections. It is very likely most of these graphs could be made quicker and more elegantly by hand drawing, provided the technologist has a small amount of artistic ability, but rapid modification of existing graphs or tables that may be necessitated by input of new data or corrections of old data is the kind of operation at which computers excel.

More complex graphs, such as response surface charts, fishnet graphs, and contour plots, involve somewhat different considerations (see, for example Macintyre 1986). Plotting the large amount of data required and orienting the diagrams by hand can be very tedious. Gaining understanding of the software and adapting it to your particular problem can also be tedious, but, when a large number of these graphs have to be made in a continuing project, it is no doubt best to develop a method for drawing them by computer. Producing figures suitably elegant and/or stylized for publication in a refereed publication can be a problem, though many computer-drawn figures can be found in today's scientific journals.

It is difficult to choose between commercially-available programs on the basis of advertised claims. There are many commercial software packages that promise to give versatile charting and graphing capabilities to the PC user, converting raw data to high quality pictorial representations with little or no effort on the part of the operator. Some of these give disappointing results and appear to be essentially fraudulent, in that the examples shown in advertisements or as printed samples could never be obtained on the equipment for which the program is said to have been designed. Part of the problem is that very detailed and fine-featured diagrams require an enormous amount of memory just for defining the pixel array or the printed dots and a memory of this size is not common in the usual desktop computer. Furthermore, monitors and printers available to most desktop PC users do not have the capability of transforming the memory data into anything remotely resembling a halftone or good line drawing.

Advances of considerable magnitude appear to be in the offing, but at this time no recommendations will be made. If an immediate procurement is essential, check out the software on your equipment using examples of the data you intend to process. Do not depend on canned demonstrations, which may have taken thousands of man hours to produce on equipment that may not be available to you.

Project Management

Some of the difficulties associated with scheduling and organizing new product development projects can be alleviated with project management software. Several programs of this general type are being offered commercially, though few of them are specifically designed for technical projects, most being better suited for planning marketing, accounting, purchasing, distribution, or other non-technical administration.

Project management software can help managers identify what has to be done and decide what resources are necessary to accomplish the goals. Some of the programs can relatively quickly prepare graphic presentations that show the status of various operations and alert the manager to potential problems. But, none of them can replace the expertise or intuitive understanding that a skilled manager can bring to a new product development project.

Most, if not all, of the popular programs use a scheduling technique known as the critical path method (CPM). The critical path leading to the solution of a problem is defined in terms of the physical or mental tasks that make up the project, their duration (the time required for completion), and the relationships among events.

The critical path can be graphically displayed on a network diagram or Gantt chart showing all of the steps necessary to accomplish a given objective and the time required to complete each task. Network diagrams, frequently called PERT (Performance Evaluation and Review) charts, are based on the interdependencies or relationships between project activities. They provide a visual representation of the activities, events, and accomplishments in relation to one another. They use boxes, arrows, and connecting lines to illustrate the links between predecessor and successor tasks), allowing the technologist to logically and methodically review a project both in its planning and operational stages. The network diagram does not ordinarily (or conveniently) depict the time allotted for each task – Gantt charts provide this durational dimension.

Each event on a Gantt diagram is represented by a horizontal bar whose length is determined by the time required for the activity's completion, showing the deadlines clearly and making it simple to recognize the status of each step in the project. In its usual form, the chart has across its top a calendar covering either the duration of the entire project or an important part of it, while the list of activities is displayed down the left side.

Low-end software can assist the manager in creating a time schedule including a relatively simple network diagram and Gantt charts. More elaborate and expensive programs have provisions for allocating human and material resources, compiling budgets, and monitoring the project as it develops.

Package Development

An example of a computer program that helps achieve the optimum package for a given product is the DuPont software called "Packaging Advisor." This program is actually somewhat limited in its application because it is restricted to packages utilizing the Dupont SELAR family of resins. In an example provided by the company, it is said that a technologist might ask the computer to specify a package structure with a shelf-life of one year, with SELAR OH as the barrier layer, at a cost of 7 or 8 cents per package. When these parameters have been entered into the computer, the program will first list the most inexpensive structure that meets all the requirements. Then it lists resin combinations that are less expensive, but don't quite meet the minimum specifications.

COMPUTER-ASSISTED MANUFACTURING

Computer-assisted manufacturing will likely be one of the options suggested during discussions of engineering new production lines or adapting existing equipment to the manufacturing of a new product. Although the technologist in charge of the laboratory phases of development leading up to, say, the prototype, may not be directly concerned with the design and construction of a production line, it is important to know the possibilities and the effects on product quality that may result from CAM plants.

All engineers recognize that automation existed before computers. Rather elaborate control systems were being built using concepts of closed-loop or open-loop control before electronic computers and programmable controllers were invented. Computers and programmable controllers use many of these previously derived concepts. A programmable controller consists of logic circuits that function as a closed-loop feedback system; it becomes a link in process control. According to Wolfe (1988), new graphics oriented PCs are well suited for use as the basis of automated systems. The power of many PCs now surpasses that of conventional dedicated instrument controllers and mini-computer based systems.

Cooper (1984) claimed computerization of plant processes can be justified because it permits a reduction in staff (or eliminates the need for increasing personnel) and yields more information from present staff. The main objectives when computerizing a plant are: (1) Choose programs that will suit the specific plant facilities and staff; (2) Look for programs first, then find the computer best suited for running the program; (3) Be sure the system is easy to use. Pick a user-friendly, menu-driven setup; (4) Find out if you will get good documentation. The printed material or directions should explain what is being shown in each menu. Documentation should be understandable by a first-time computer user; and (5) The supplier should have a good knowledge of the food manufacturing business, and should also have the ability to support both hardware and software.

Among the production information that can be obtained from computerized production lines are: (1) Instant recap of all products that have been ordered; (2) Production orders broken down into numbers of pounds, packages, cases, and loads; (3) Weigh lists for the mixing room, with formulas and total batch weights; (4) Raw material usage, including number of batches; (5) Packaging material usage; and (6) Cost of the day's production.

According to one engineer, VMS and OS/400 have been the most popular operating systems for developing manufacturing software applications in recent years, but they are losing popularity to DOS and UNIX. Use of OS/2 is also increasing, but it is starting from a very low base of only two or three percent.

Programmable Controllers

As pointed out by Booth (1987), automation of mixing and other manufacturing processes is best achieved by using a computer that can interface with microprocessors on the production equipment. Control of an automated mixing system can be achieved by interfacing a mainframe or a mini- or micro-computer with ingredient dispensing operations and mixer, using a programmable controller. The role of the computer is to: (1) Store recipes and ingredient tolerances that have predetermined as well as inventory data for usage records and re-order purposes; (2) Store and print out batch numbers, batch weights, individual ingredient weights, and batch temperatures; (3) Make changes in recipes, mix times, and temperatures; and (4) Control and monitor the equipment. Some of the earliest advances in computer control of food production appear to have taken place in large bakeries, and the following discussion will relate mostly to bakery applications.

The programmable controller performs the simple logic and sequencing operations required by the system, receiving computer commands and passing them on after making the necessary decisions as to destination, timing, etc. The links between the controller and the process are sensors and activators. It interfaces with the hardware on the production equipment, opening and closing valves, and actuating the other mechanical devices. It can be programmed using standard relay ladder logic, familiar to most electricians and maintenance personnel.

Programmable controllers can be applied to existing equipment, i.e., retrofitted. If simple control circuits are needed to perform many rapid cycles, a programmable controller may be justified. Most mechanical control devices and switches will wear out after a few million cycles, which can be reached in a few weeks in modern automated plants. An example is a bulk bun packer having controls that may cycle 200,000 times per week. Solid state control devices are not limited by mechanical wear and tear. Cost of a programmable controller can be quickly paid back by eliminating the down time needed for replacing mechanical switches.

Programmable controllers are flexible and can be adapted to a different application if their original use becomes obsolete. They are simply programmed to perform the new function. This flexibility is also helpful when add-ons are needed on the original machine. Generally, the controller can be programmed to encompass the additional operations.

McCaffery (1986) described a high performance, programmable controller applied to a bakery production line. As part of its functions of controlling and monitoring all of the key equipment in the plant, the controller can send information to, and receive it from, a personal computer. Components of the system include two high-capacity, advanced-technology personal computers, one located in the production control center and the other in the chief engineer's office. There are ten remote input/output locations, each of which includes input cards, output cards, a rack to plug these cards into, and a serial interface module. The latter device takes the place of the microprocessor in a small independent programmable controller. Its main purpose is to provide a translating device to take signals from the input/output rack and route them through a specially designed coaxial cable that allows communication to remote locations up to 1,000 ft from the controller so that plant layout is not limited by system hardware.

Terminals for operator sending and receiving are located in both the QA lab and the floor control stations. The PC in the engineer's office is mainly concerned with maintenance functions, while the PC in production control primarily logs and records the operating characteristics of the plant (e.g., time, temperature, and humidity). The system can scan all points in 75 milliseconds. Human errors are minimized, according to McCaffery.

Examples of equipment that can benefit from programmable controllers are pan handling conveyors and bulk flour handling systems. A pan handling conveyor can be improved by using a programmable controller that can count the pans in a predetermined area and make adjustments based on other factors in the system. This reduces the number of persons needed to ensure that the correct number of pans are on the proper section of the conveyor at a given time and gives a steadier and more uniform pan flow. Bulk handling systems for flour will require accurate timing and precise control for satisfactory performance. With conventional controls, extra time must be allowed to ensure that the speed of the system is never exceeded. If this margin is not allowed, standard controls will begin to vary the set points as wear occurs, and stoppages may result. A reliable programmable controller will not need the safety margins in timing and can make it possible to transport many more pounds of flour or other ingredient per hour (Bailey 1985).

Automated dough temperature systems are computer controlled. A dual output temperature controller/indicator works in conjunction with a time on the refrigeration system. One set point controls the final dough temperature while the second set point is adjusted to an upper temperature limit considered to represent an unacceptable dough. If the normal mix time is completed at a time when the dough temperature exceeds the upper limit, the mixer will continue running in slow speed with full refrigeration until the dough cools sufficiently to meet specifications (Booth 1987).

From the monitor, visual indications can be obtained of the status of make-up equipment, status of the safety devices, and any failures of components. The advantages of a computerized automated mixing system, as enumerated by Booth, are: (1) labor savings – no one is in attendance at the mixer; (2) reasonable cost; (3) inventory control – provides an accurate, constantly updated record of materials used and materials remaining in stock; (4) product consistency – ensures constant recipe balance regardless of batch size; (5) no formulation change – existing recipes can be kept when a switch to automated mixing is made; and (6) instant recognition of problems such as tripped safeties and blown fuses simplifies trouble shooting.

Computer-controlled ingredient dispensing allows proportional batching. For example, the amounts of all other ingredients can be based on the amount of flour delivered to the mixer, thus retaining integrity of the ingredient proportions even though flour weight may have been adjusted because of moisture variations, etc. In a typical system, the supervisor selects the type of product to be made by pressing a button labeled with the appropriate name. The product category, a list of ingredients, or possible varieties then appear on the display screen of the instrument panel. Ingredient feeding is then actuated by the operator. As flour is loaded into the scale hopper, the actual weight will be displayed in one column and percentage variance in another column. If the actual amount fed is outside a preset variance limit, an alarm will signal the need for supervisory action. Assuming the flour weight is within acceptable limits, the remaining ingredients will be fed into the scale hopper in proportions adjusted to the actual flour weight. When the preceding batch has completed its mixing cycle and the bowl has been emptied, the total ingredient mix will be dumped into the bowl.

Sensors

Information about the characteristics of the object of interest, perhaps a cookie, a peanut, a sugar solution, or a mixer load of dough, must be obtained by examining that item before the computer can do its job of analyzing and reporting the information that will be used by someone to evaluate the item or adjust the process acting upon it. This information is obtained by sensors, of which there are many kinds. Only a few of the types can be discussed in this article.

Sensors are available for detecting temperature, pH, weight, distance, color, pressure, and many other attributes of samples and environments. They provide the computer with digitalized electrical signals that can be processed into a form useful for further handling or for reporting to a human observer. For example, a range or distance detector emits light or sound waves that are reflected from an object and then detected by the sensor. The time required for the light or sound to make the round trip is calculated and transformed by the computer to a distance stated in common units such as inches or centimeters.

Of somewhat similar function are proximity sensors, which can determine the presence of an object within their sensing area without the necessity for touching the item. They can be used to count pans, rolls, or almost any other object moving along a conveyor. Electromagnetic sensors are used to determine the presence or absence of metallic substances. Temperature sensors include contact types such as thermistors and thermocouples and non-contact types that measure radiation emitted by hot objects.

Electromagnetic sensors are used to determine the presence or absence of metallic substances. Temperature sensors include contact types such as thermistors and thermocouples and non-contact types that measure irradiation emitted by hot objects. Chemical sensors convert chemical information into electronic signals. Typically, they are miniaturized devices that perform analytical chemical functions closer to the point of interest, more frequently, and at lower cost than conventional laboratory instruments (Goldman 1988). They are useful in process control and environmental monitoring, as well as in other applications, such as biosensory measurements. A chemical sensor is usually composed of a coating, a sensing substrate, a means of transmitting information to and from the remote sensor, electrical or electro-optic hardware to analyze the signals, and a computer for data reduction. Most of them can be classified as optical, acoustical, or electrochemical. Ion-selective electrodes and pH electrodes are chemical sensors that are useful in process control and lab tests.

Further information on this subject can be obtained from the article by Caro and Morgan (1991) and other papers listed in the biblography.

Inspection Applications

Computers can provide important advantages to inspection systems by processing data obtained on the production line. To fully utilize the capabilities of computers in such applications, it is necessary to eliminate, as much as possible, the human element. If the computer is used only to collate and analyze measurements fed into the system by people who are looking at, weighing, and measuring products, maximum efficiencies cannot be obtained. The best results can be had when measurements are made by automated inspection systems. Such systems require specialized sensors and computers with software that can interpret them.

Egan (1988) quotes the following definition of an automated inspection system from a standard issued by Automated Vision Association: "Devices used for optical non-contact sensing to receive and interpret automatically an image of a real scene in order to obtain information and/or to control machines or processes."

Automated inspection offers speed and objectivity unattainable with manual inspection. As line speeds increase, it ultimately becomes impossible for human observers to make the necessary differentiation between separate objects on the conveyor, but there is no practical limit to the speeds attainable with machine vision or automated inspection.

The color of most food products can be determined electronically by measuring light reflectance at one or more wavelengths. Measuring the entire spectrum of reflected visible light is neither necessary nor practical. For example, caramelization or non-enzymic browning reactions, such as occur in frying foods, are best measured at the longer wavelengths; there is generally a decrease in reflectance in the blue region, some change in the green, and a considerable drop in the red. The latter change frequently continues into the infrared portion of the spectrum (Coatney 1986).

In detecting the process of crust coloration during baking of, for example, loaf bread, there is little change in the blue region of the spectrum, considerable change in green and red reflectance, and very little change in the infrared. Coatney described the use of the Agtron CA II color sensor in quantifying the appearance of potato chips. This device was developed as an intelligent color sensor capable of full communication with a process control computer system.

Shape and dimensions can be measured by optical sensors. Egan described the need for measuring the height of crackers as they exit the oven. The most important datum is the maximum height of the crackers. Crackers that have excessive thickness will not fit into the box when the predetermined number of units is assembled in the quarter-pound stack or "slug." If the height begins to increase, the computer can signal the sheeting rolls, causing them to make the dough web thinner. It is necessary to measure the height of the product relative to the conveyor since the conveyor itself varies in distance from the optical sensor depending on irregularities in the steel belt, bouncing, etc. Small changes in height (less than 0.003 inch) must be detected, and the sensor must be able to differentiate between bouncing of the conveyor and a true increase in cracker thickness.

Modified stock designs of nuclear magnetic resonance (NMR) spectrometers have been used as on-line process control analyzers, but they proved to be too difficult for plant operators to calibrate and they were difficult to maintain. To eliminate some of these problems and minimize the time required to train plant operators in the use of NMR, the spectrometers were designed to include several self-calibrating and self-adjusting features (Pearson *et al.* 1987). Before an NMR analyzer can be used for on-line process control, sampling and control system hardware must also be added. Moisture and oil contents of ground corn and moisture content of tempered wheat were satisfactorily determined by the modified equipment.

A paper by Lawson (1984), which dealt with computer applications in the cookie and cracker industry, especially as that industry operates in the UK, describes the use of programmable logic devices to control ovens. He states that information that can be acquired, stored, and distributed by an oven PLC includes fault finding, maintenance procedures, fuel rate, output of product, efficiency of oven, heat exchanger temperatures, band temperature, flue analysis and volume, biscuit dimensions, state of oven, band tension, band wander, and data from other machines such as the forming machine or packer can be entered into the memory — one set for each formula.

Robot Operators

Robots have been defined as re-programmable, multi-functional manipulators designed to move some kind of tool, material, part, or semi-finished product through variable programmed motions for the performance of a variety of tasks. It would no doubt be generally agreed that the characteristic that distinguishes a robot from other types of machinery performing repetitive motions, is that a robot has the ability to vary its performance according to the operator's setting of adjustable mechanisms or, alternatively, modifies its movements according to signals received from a built-in intelligence or from a programmable memory in a central processing unit (Spooner 1985). From this definition it can be understood that, to be called a robot, a device does not necessarily have to be a computer; it does not even have to be connected to a computer, but it is a fact that virtually all robots being designed today are integral parts of systems containing computers.

Robots have been classified as the following types: limited sequence playback with point-to-point control, and playback with continuous path control. The first type uses a combination of mechanical stops and limit switches to control the movements. These robots have no memory unit and cannot be programmed to perform their task but have to be set up in the same way an automated machine would be adjusted. The second type obtains positional control with a servo system that electrically detects and corrects errors. Memory units must be added to electronically retain all of the position commands before a true robot can be said to exist.

Although the jointed metal arms and plier-like fingers seen in so many TV specials on robots may not be visible in a modern plant, the automated production lines that have been operating for years have most of the basic attributes of robots. According to Spooner, however, an industrial robot provides some additional benefits over automated machinery. These are:

(1) Availability – it may be possible to get an off-the-shelf (i.e., stock) industrial robot adaptable to an existing or planned production system, avoiding the extensive line design and construction of special equipment that might otherwise be required.

(2) Adjusting and debugging – when conventional automated production equipment is installed, indexing devices, solenoids, actuators, and clamping mechanisms must be adjusted on the plant floor after delivery, while the robot can be pre-programmed and put into use quickly after it is received.

(3) Resistance to obsolescence – robots provide a form of automation less subject to obsolescence. A robot need not be product- or operation-limited. It can be re-programmed to meet changing requirements.

Some expected advances in robot design that will make robots more useful in industrial applications are complex optical (visual) recognition and guidance capabilities, sensing and identifying tactile and chemical signals, more flexibility and versatility in the movements of the product-contacting element, response to voice commands, and total diagnostic fault tracing.

Packaging

Virtually every type of packaging machine can be provided with a control system that will insure uniformity of weights, provide efficient channeling of products to the machines, and monitor package characteristics. Topken (1984) described the history of automating and computerizing packaging lines, and the following discussion is based, in part, on his paper.

Early control systems contained operator controls such as start and stop buttons, electrical timing devices with one or more limit switches, electromechanical devices such as solenoids, and the control logic to decide what the electromechanical devices were to do in response to input signals. Most logic operations were performed by relays.

Due to the demand for more automation, sensors were added to the machines and more output devices were added to the system. The output devices included additional electromechanical equipment, enunciators of various descriptions, and production monitors. With more components and electrical connections, reliability became a problem. Changes were more difficult to make because of the complexity of the system.

The programmable controller made the task of designing complex control systems much simpler and less expensive — many of the moving parts and much of the hard wiring were eliminated. As other forms of microprocessor controls became available, they were added to the system. Timing limit switches were replaced with programmable limit switches, which were more reliable. Servo-motors and stepping motors enabled the designers to eliminate bulk mechanical drives. For example, timing controls based on microprocessors maintained consistent patterns of glue on cartons even when the speed of the machine changed. This feature was needed so the speed of machines could be changed automatically for more efficient operation under conditions of varying product at the infeed.

Sophisticated enunciator systems were designed to help the operator maintain efficient machine functioning. Digital readouts became popular. One result of all these changes was that the operator control station became much more complicated. Elements that previously operated independently are now tied together with a central microprocessor or group of microprocessors. This system accepts inputs from both sensors and operator controls. Custom designed enunciator systems are replaced by cathode ray displays. Complexity of operator controls is dramatically reduced, leaving only four pushbuttons for routine operation. To ease the task of the operators and avoid errors from information overkill, an integrated system should also allow improvements to be achieved through software changes.

GLOSSARY

For the convenience of readers who are considering the integration of computers into some part of their product development program, the author has included a very brief glossary of technical terms. The following list includes many words and phrases from the arcane language used by persons in computer sales and service, writers on computer terminology (LIMS, LANs, and allied subjects), and hackers. A considerable number of desktop publishing terms are included because of the importance of this computer application to most laboratory workers.

access server – a multimodem device for handling multiple simultaneous dial-in connections to a network.

acoustic coupler – device that enables computer signals to be transmitted through conventional telephone lines

artificial intelligence – computer systems that simulate human intelligence by devising solutions to problems in the absence of specific programs detailing the sequence of operations necessary to solve the problem. The computer can "learn" how to do complex tasks without the guidance of a programmer or software specific to the task.

ASCII – a standardized code for representing characters as binary numbers; it is used on most computers and printers.

bar code – a pattern of wide and narrow parallel bars printed on paper or other material, readable by computerized vision systems.

batch – a group of instructions and data that will always be automatically entered into active memory when a certain program is called for.

baud – a measure that indicates the speed of data transmitted by, e.g., a modem; 1 baud equals 1 bit per second.

bit – abbreviation for "binary number," either 1 or 0 in computer language. A bit is, physically, a memory cell, the presence or absence of magnetization at a specific location on a disc or tape, or a pulse of electricity traveling through a circuit.

bridge – interconnection device for linking LANs and signaling systems such as Ethernet, and Token-Ring.

brouter – device combining the functions of a bridge and a router,

buffer – short-term storage system allowing data to be accumulated when the receiving device (such as a printer) uses data slower than it receives it; sometimes called a "spooler."

bus – a trunk cable; also, a network design based on a single bus that two or more stations can be tied into for receiving the same message at the same time.

bus mastering adapter – network adapter with circuitry capable of taking control of the PC's expansion bus, relieving the CPU of the PC of much of the network overhead.

cache memory– a small high-speed memory for the temporary storage of information, usually placed between a relatively slow large memory (as on a disk) and a fast central processing unit.

CAD/CAM – computer-aided design/computer-aided manufacturing.

CIM – computer-integrated manufacturing

coaxial cable – a shielded single wire conductor.

command – a series of characters (a "string") that directs the computer to perform a specific operation.

concentrator – central meeting point of a star-wiring scheme.

CSMA/CD – media-sharing method used by Ethernet that allows any station to transmit if the cable is idle.

database – a collection of interrelated files stored together and available for multiple applications.

data switch – external box that allows PCs to make connections with other PCs and peripherals.

dedicated server – a computer that performs specialized network tasks.

distributed file system – the part of a network operating system that allows one computer to use the files and peripherals of another.

dpi – dots per inch, a common linear measure of resolution or definition in monitors and printers.

emulator – a program that makes one computer work like another computer with which it is normally incompatible.

enhanced parallel port – improvement that greatly increases information transfer speed.

Ethernet – a network cable and access protocol, the most common UNIX network scheme, developed by Xerox.

external adapter – external conversion device that allows the parallel printer port of a PC to be connected to a network.

fiber-optic cable – cable containing a set of thin strands of glass or plastic that permits very high rates of transmission.

file – a collection of related records treated as a basic unit of storage.

file server – the disk storage and associated controlling hardware and software that are attached to a LAN so that devices using the network can share storage space.

font – a complete collection of letters, punctuation marks, numbers, and special characters in a single typeface, all having the same weight (e.g., Roman or bold) and posture (e.g., upright or italic). For bit-mapped fonts, there will also be only a single point size.

font family – a set of fonts based on a single typeface but containing several members having different weights and sizes.

frame – a packet on a Token-Ring network.

full-duplex – serial communications allowing simultaneous data flow in both directions.

gateway – allows communication between two networks running incompatible protocols.

half-duplex – serial communications that allow transmission in two directions, but not at the same time.

hub – a wiring center that is the junction of a star wiring topology.

input/output device – equipment connected to a computer for putting in or taking out data.

intelligent terminal – a terminal that can function either as part of a network or as a self-contained computer.

justified – lines of type that are flush at both the left and right edges, a feature available in all currently offered desktop publishing software.

kerning – to place two letters closer together than customary in order to create visually consistent spacing; often, but not always, accomplished by actually joining the two letters (called a ligature). The number of kerned pairs available in a typeface is one indication of the sophistication of the software font package.

LAN – local area network of computers, peripherals, and software interconnected so they can share disk storage space and printer resources.

LIM – laboratory information system involving the integration of computers into data collecting, storing, sorting, and sharing.

locking – protecting shared data by preventing simultaneous access by a second program (or limits access to read-only).

mainframe – large-scale computer system housing comprehensive software and several peripherals.

media – wires, cables, and connectors that link the network and carry information. Also, disks for storing data (i.e., memory).

media filter – device acting as a switch between UTP and STP cables in a Token-Ring network.

modem – device that converts data into and out of high-frequency communications signals for transmittal over telephone circuits.

NetBIOS – network basic input/output system, software for linking a network operating system with specific hardware.

network adapter card – (NAC) the physical and logical link between the computer and the network.

node – each PC, workstation, and server in a network system is called a "node" and has a unique address.
null modem – a set of connectors allowing two computers to communicate as though they were transmitting through modems.

OSI – a model for networks that was developed by the International Standards Organization.

packet – the actual message (including addressing and other data) passed through a network; called "frame" in some systems.
parallel transmission – a methhod whereby a number of bits are sent simultaneously over separate lines, usually unidirectionally.
password – a secret character string that acts as an electronic key giving access to a computer system.
peer-to-peer – network architecture in which each user's PC operates as a server rather than accessing centralized file or print servers.
pica – traditionally, metal type approximately one-sixth of an inch in size; now refers to a measure of type equal to 12 points (q.v.), or to a typewriter type size. Some desktop publishing software unfortunately makes use of this obsolete measuring system.
pixel – a single dot marking a unique location on a display screen.
point – a measurement used in typography, and to some extent in desktop publishing, equivalent to 0.0138 inch.
PMIS – process manufacturing information system.
protocol – in computer terminology, a formal set of conventions governing the formatting and timing of message exchange between two communicating systems.

realtime – an electronic operation that occurs in the same timeframe as its real-world counterpart.
redirector – software module in a network workstation that captures application programs' requests for files and equipment sharing services and redirects them out into the network.
repeater – the simplest LAN interconnection device, extends the length of the network media and, if buffered, amplifies and regenerates signals.
RISC – describes a type of computer architecture (Reduced Instruction Set Computer) that performs a limited number of operations, thereby speeding performance of its overall task. The basic concept is that most computer programs require only a few instructions and, if those essential instructions can be made to occur quicker, overall performance will be accelerated.
router – interconnection device that sends defined protocols on to the correct links.

scalable font – a means of forming a group of fonts from a single design or pattern (transformed into electronic signals, of course) that defines only the outline of each letter; thus a 6-point printout would base its printer input on the same basic information as a 16-point printout, with uniform changes in size throughout. Examples are Adobe PostScript and Apple's True Type; recent issues from IBM and Microsoft use this software strategy.

SCSI – "small computer system interface." Refers to a computer system, or design, that includes a universal expansion socket permitting users to connect multiple add-on devices such as hard disk drives, scanners, or printers without opening the computer's cover. Pronounced "scuzzy."

serial transmission – method in which information bits are sent sequentially on a single data communication channel.

star – wiring topology for networks that are configured like the hubs and spokes of a wheel.

synchronous terminal – describes a data terminal operating at a fixed rate with synchronized transmitter and receiver.

TELNET – terminal emulation and communication protocol.

token – a special message that allows some action to be taken.

token passing – access protocol in which a token circulates among the network modes, giving them permission to transmit.

Token-Ring – IBM's network architecture in which a token, passed from node to node, gives a node permission to send a frame onto the ring.

topology – electrical (or physical topology) is a map of the network describing how the wires or cables are laid out; logical topology describes how messages flow.

twisted-pair wiring – a ground wire and a signal wire twisted together at six turns per inch to provide a self-shielding cable, as used to connect nodes in some networks.

type family – a set of typeface designs that are all variations of one basic style or pattern of an alphabet. For example, many type families consist of roman, italic, and bold variations, while more sophisticated versions will include modifications in width (condensed or extended) and weight (light to extra bold).

weight – in desktop publishing and typography, weight refers to the ratio between a letter's solid strokes and its open counters; "heavyweight" if the strokes are thicker than the basic pattern and "lightweight" if the strokes are thinner.

workstation – intelligent terminal system, including the application software for it.

write-back cache – in such systems, memory-write as well as memory-read data are stored in the fast cache; data that the CPU writes to memory goes initially only to fast cache memory, avoiding wait states, and the cache controller moves that data to slower main memory only if it is about to be overwritten by some other memory operation.
write-through – places only the memory-reads; a variation is "posted-write," it can delay transfers to memory with a small write-buffer, the contents of which go to main memory only when the CPU moves on to non-memory operations.

X.25 – a protocol for long-distance packet-switched transmissions that allows networks with different protocols to communicate.

BIBLIOGRAPHY

ALY, N. A. 1989. A survey on the use of computer-integrated manufacturing in food processing companies. Food Technology *43*, No. 3, 82, 84-85, 87.

ANON. 1989. Baking the best bun the automated way. Baking & Snack Systems *11*, No. 5, 7, 9.

ARNOLD, G. R. 1989. Computer integrated manufacturing. Proc. Am. Soc. Bakery Engrs. *1989*, 50-58.

ARNOLD, G. R. 1990. More on CIM. Proc. Am. Soc. Bakery Engrs. *1990*, 169-179.

BAILEY, N. A. 1985. Programmable controllers and existing equipment. Proc. Am. Soc. Bakery Engrs. *1985*, 190-195.

BLASER, W. W., RUHL, H. D., and BREDEWEG, R. A. 1989. The challenge of on-line analytical instrumentation. American Laboratory *21*, No. 1, 69-70, 72.

BOCKELMAN, J. 1990. Quality control brings sophistication to at-line testing. Baking & Snack Systems *12*, No. 6, 20-24.

BOOTH, M. D. 1987. Automated high volume dough mixing. Proc. Am. Soc. Bakery Engrs. *1987*, 132-144.

BUSH, P., and LAHVIC, R. 1989. Pepperidge Farm's "Project Freshness." Prepared Foods *158*, No. 2, 122-124.

CARO, R. H., and MORGAN, W. E. 1991. Trends in process control and instrumentation. Food Technology *45*, 62, 64-66.

COOPER, E. L.. HICE, R., and RAHN, P. D. 1992. The implementation and integration of 2nd generation LIMS. Scientific Computing *8*, No. 3, 45-49.

COOPER, M. C. 1984. Computers for the baking industry. Proc. Am. Soc. Bakery Engrs. *1984*, 99-105.

CURTIS, J. W. 1987. Robotics and artificial intelligence for the food industry. Food Technology *41*, No. 12, 63-64.

DAVIS, A., BELTON, G., and MATTHEWS, D. 1988. Interfacing laboratory instruments to an LIMS computer system. Am. Lab. *20*, No. 9, 34, 36-41

DEMING, R., and YOUNG, L. A. 1989. Laboratory computer networking. Scientific Computing & Automation *5*, No. 4, 21-25.

FINDLAY, C. J., GULLETT, E. A., and GENNER, D. 1986. Integrated computerized sensory analysis. J. Sensory Studies *1*, 307-314.

FLOROS, J. D., and CHINNAN, M. S. 1988. Computer graphics-assisted optimization for product and process development. Food Technol. *42*, No. 2, 72, 74-78.

FOWLER, B. 1988. Integrated laboratory automation. Am. Lab. *20*, No. 9, 62, 64-68.

FUSARO, D. 1993. What operating systems are engineers using? Control *6*, No. 4, 18-19, 21-22, 25.

GOLDMAN, D. S. 1988. Chemical sensor development for process and environmental monitoring. Am. Lab. *20*, No. 11, 102, 104-109.

GORTON, L. 1988A. Computers integrate packaging, distribution. Baking & Snack Systems *10*, No. 4, 6-9.

GORTON, L. 1988B. CIM put in control at new Danish plant. Baking & Snack Sytems *10*, No. 6, 6-8, 10-13.

GORTON, L. 1989. From mechanization to true automation. Baking & Snack Systems *11*, No. 5, 24, 26, 28-29.

HAAS, D. J., and MURPHY, L. 1989. Bar codes in the laboratory. Scientific Computing & Automation *5*, No. 10, 49-55.

HATFIELD, P. 1993. Saving money with open systems. Control *6*, No. 4, 34-36.

HETTICH, G. 1989. The basics of food automation. Prepared Foods *159*, No. 4, 132.

HIGHAM, J. D. 1991. Automation in shipping and packaging. Proc. Am. Soc. Bakery Engrs. *1991*, 150-158.

HUTCHISON, K. M. 1989. Rapid real-time monitoring and direct data storage techniques. Am. Lab. *21*, No. 1, 102.

JACKEL, S. 1984. Microcomputers, microprocessors, and robots in bakery operations. Cereal Foods World *29*, 258.

JOGLEKAR, A. M., and MAY, A. T. 1990. Accelerating product development by computer-aided design of experiments. Cereal Foods World *35*, 1187-1192, 1197-1200.

JONES, G. Y. 1985. Introduction to programmable controllers. Proc. Am. Soc. Bakery Engineers *1985*, 177-181.

JOYCE, R. 1989. Advances in PC-based data acquisition. Am. Lab. *21*, No.6, 48, 50, 52-54.

LEONARD, L. 1980. The big computer bake-off at Sara Lee. Prodn. Eng. *27*, No. 1, 39-43.

LYSAKOWSKI, R., JR. 1992. Toward laboratory automation system standards. Am. Lab. *24*, No. 3, 34, 36, 38, 40-42.

MACINTYRE, F. 1986. Generation of fishnet graphs and character-mapped contour plots. Am. Lab. *18*, No. 9, 21-27.

MILLER, L. 1991. Computerized quality control system. Food Technology *45*, 102

MISROCH, M. 1988. Planning and evaluating a laboratory information system. Am. Lab. *20*, No. 9, 28, 30, 32.

NEWMAN, F. P. 1991. Fourteen shortcomings of current control technology. Control *4*, No. 4, 53-55.

PASDIRTZ, G. 1990. Laboratory automation, interface with the future. Hazelton Science Newsletter *34*, 1-5.

RYAN, J. 1991. LIMS aid environmental labs. R&D Magazine *33*, No. 8, 59-60.

SCHULTZ, T. M. 1987. Computer uses for plant maintenance. Proc. Am. Soc. Bakery Engrs. *1987*, 162-168.

SELLERS, C., and POPPITI, J. 1989. LIMS the QA equation. Environmental Lab *1*, No. 8, 18-24.

SHOOK, P 1993. PCs finally getting plant-floor respect. Control *6*, No. 4, 38-39, 41.

SPERBER, R. 1991. PLCs cure downtime with diagnostics. Food Proc. *52*, No. 10, 78, 80.

SPERBER, R. 1993. New technology spices approach to automation. Control *6*, No. 4, 74-76, 78, 82.

SPOONER, T. F. 1985. Robotics. Proc. Am. Soc. Bakery Engrs. *1985*, 168-177.

SPOONER, T. F. 1989. Sensors emerge for artificial intelligence uses. Baking & Snack Systems *11*, No. 6, 21-24.

SPOONER, T. F. 1991. Bridging the expertise gap. Baking & Snack *13*, No. 7, 22-25.

STAHL, D. C. 1991. Process development is easy with the right program. R&D Mag. (Nov.), 73-74.

VAIHIANATHAN, R. 1985. Using sensors in bakery production. Bakery Production Marketing *20*, No. 2, 110, 112.

WALULEK, S. C. 1985. Programmable controllers and energy. Proc. Am. Soc. Bakery Engineers *1985*, 181-189.

WEISS, M. D. 1991. FT-IR moves out of the lab. Control *4*, No. 9, 36, 38, 40.

WILSON, D. 1985. Retrofit equipment with electronics for competitive edge. Bakery Production Marketing *20*, No. 2, 106, 108, 110.

WOLFE, R. 1988. New PCs are ideal platforms for automation. Scientific Computing Automation *4*, No. 6, 18, 20-22.

YACU, W. 1990. Process instrumentation and control in food extruders. Cereal Foods World *35*, 919-926.

YAZBAK, G. 1991. Fiberoptic sensors solve measuring problems. Food Technol. *45*, 76-78.

ZIPPERMAN, W. 1991. Software tools help bridge the CIM gap. Control *4*, No. 11, 52-54.

chapter thirteen

PROTECTING INNOVATIONS

INTRODUCTION

An important part of the product development process is the step or practice of protecting innovations so that competitive copying can be prevented or at least delayed. If you develop and market a successful product (or even a semi-successful product) some of your competitors will decide to imitate it. They will assume your company has completed, and were satisfied with the results of, evaluations of production feasibility, consumer acceptability, economic viability, and regulatory compliance of the new product. They will further assume that they can safely omit many of these expensive and time-consuming steps if they decide to market an identical, or very similar, product. They will track wholesale and retail movement of your product through the type of commercial distribution you have chosen, thus determining whether there is a stable repeat demand for your product without the need for a large prior investment in market and sales tests. Fortunately, there are a number of legal measures that you can take to defend your priority – at least for a time.

There are four procedures that have been extensively used in the past for isolating new products and new lines of products from competitive pressures. Three of these – patenting inventions, registering trademarks, and copyrighting text and the like – are formal methods requiring the exposure of your ideas in fully understandable form to all interested parties during the course of required registration procedures, while the fourth takes the opposite approach of concealing as much as possible of the idea. All of these approaches are based on rights existing in the common law, much modified by statutory changes and prior court decisions. Trademarks and copyrighting, particularly, are used primarily for purposes other than protecting new products, but these broader applications will not be discussed in this book.

A fifth type of protection that has recently assumed added importance is litigating on the basis of alleged competitive copying of "trade dress." This concept is at present in the formative stages. Unlike the other four measures, it requires neither formal recording of supposed innovations nor deliberate suppression of public exposure of your ideas. Up to this time, it has been used mostly to obtain relief from competitive copying of rather broad areas of business practice. How it can be applied to the protection of new products has yet to be determined by the courts.

You should protect your ideas and your work in as many ways as possible so you can justifiably profit from the expenses involved in developing, producing, and marketing the new product. For maximum effectiveness in protecting your innovation, it is a good idea to bring your legal advisers into the picture at an early stage of the development – preferably as soon as you are reasonably sure that a true innovation is in prospect. The following discussion is not intended to substitute for an attorney's counsel, but it will point out some of the simpler methods that can and should be used throughout the development process to give your attorney some basis for his efforts.

TRADEMARKS

Particularly if your new product or process is going to be offered directly to consumers (but in other situations as well), you will probably find it advisable to devise for it a name or symbol that can be restricted to use by your company. Such specialized names and symbols are called "Trademarks" and they can be registered in most, perhaps all, states and territories of the U.S., in the Patent and Trademark Office of the U.S. Department of Commerce, and in some foreign countries. The basic features of trademark registration on the Federal level will be covered in the following comments, many of which are taken directly, or with slight paraphrasing, from various publications of the Patent and Trademark Office, U.S. Department of Commerce (Anon. 1989, etc.)

A trademark may be a word, symbol, design, slogan, color, or even a distinctive sound that distinguishes the goods or services of one firm from those of another. A trademark appears on products or packaging while a "servicemark" is usually a word or design used in advertising to identify a person or company selling services. Trademark rights arise from use of the mark; registration is not essential to develop property rights in a mark but registration does result in a number of practical and legal advantages in protecting your property. Prior to its registration, users of a mark will customarily identify it with the superscript "TM" (or "SM" for a service mark); after registration, the encircled "R" is used.

A trademark must be in use before it can be claimed, though the use need not be extensive in time or geography. A mere idea that has not been put into use cannot be registered, though a bona fide intent to use it in trade together with certain actions showing this intent may give a basis for protection.

Benefits of Registration

Among the benefits that may accrue to the trademark user from registering the mark with the Patent and Trademark Office are:

1. You have a right to sue in Federal court for trademark infringement.
2. You can recover lost profits, damages, and costs from an infringement action filed in a Federal court infringement action and may possibly obtain treble damages and attorney's fees.

3. Registration is constructive notice to all others that you claim ownership. This eliminates the use of a good faith defense by a violator who acts after the date of registration.

4. You can deposit the trademark with Customs in order to halt the importation of goods bearing an infringing mark.

5. Registration is prima facie evidence of validity of the mark, registrant's ownership of the mark, and registrant's exclusive right to use the mark in connection with the goods or services specified in the certificate.

6. Once the registration has been in effect for five years, there are limited grounds for attacking it.

7. There is the possibility of criminal penalties for counterfeiting a registered trademark.

8. The U. S. registration provides a basis for applying for similar protection in foreign countries.

Registration Procedure

Generally speaking, a trademark or servicemark does not cover the whole universe of goods and services. The Federal government provides a standard list of 34 classes of goods and 8 classes of services. An application to register a trademark must indicate which of these classes are claimed. The ones of most interest to readers of this book will probably be 29 and 30 – which include most types of regular foods; 32 – which includes many kinds of alcoholic and non-alcoholic beverages; 31 – which includes fresh fruits and vegetables, grains not covered specifically in other classes, animal feed, malt, and certain other food and nonfood products; and 5 – which includes infants' and invalids' foods, among other products. The descriptions in their entirety for classes 29 and 30 are:

Class 29: Meats, fish, poultry and game; meat extracts; preserved, dried, and cooked fruits and vegetables; jellies, jams, and preserves; eggs; milk and other dairy products; edible oils and fats; pickles.

Class 30: Coffee, coffee substitutes, tea, and cocoa; sugar; rice; tapioca; flour and preparations made from cereals; bread, biscuits, cakes, pastry, and confectionery; ices; honey; treacle; yeast and baking powder; condiments such as salt, mustard, pepper, spices, vinegar, and sauces; and ice.

It is possible to claim more than one class for a single trademark. Different owners may claim the same trademark for different classes, for example, one company may claim the word "Plymouth" for vehicles and another company may claim the same word for soup. Difficulties in legal interpretation may arise when one company owns a trademark for e.g., automobiles and another company attempts to register the same word, symbol, or sound for e.g., bicycles.

Prior to filing for registration, the claimant must have a search made for existing registrations of marks resembling the claimed mark. Obtaining this information, and analyzing it once it is in hand, is a fairly complex matter and should by handled by an attorney specializing in trademarks. From the time the first search is made to the time the certificate is received can be 18 months, the duration in each particular case depending on the diligence of the lawyer and the number and types of conflicts uncovered in the search. The government does not require that a lawyer be employed to file the certificate.

Computerized searching techniques have greatly reduced the time and cost required for reviewing the trademark registration files of the states and of the Patent and Trademark Office. These databases contain probably in excess of one and a quarter million trademarks, many of them being inactive. Private companies have been formed for the specific purpose of searching these databases. A quick and relatively inexpensive review of existing trademarks can be done for anyone; you don't have to be a lawyer to make use of the services. Two of the best known databases for this purpose are TrademarkScan (on Dialog) and CompuMark.

Thomson & Thomson, a trademark and copyright research organization, announced (1993) the availability of a CD-ROM product line with the following new features: (1) subscribers may now simultaneously search across all discs in the Federal database, (2) full histories of your sessions may be saved and printed from the Federal and state CD-ROMs, and (3) the number of discs has been reduced by data compression techniques. The phone number of this firm is (800) 692-8833, and its address is Thomson & Thomson's Online/CD-ROM Services Department, 500 Victory road, North Quincy, MA 02171.

A trademark application consists of (1) a written application form, (2) a drawing of the mark, (3) the required filing fee, and, if the application is filed based upon prior use of the mark in commerce, (4) three specimens showing actual use of the mark on or in connection with the goods or services. A separate application must be filed for each mark for which registration is requested.

There are currently two standard forms in use for the initial submittal to the Patent and Trademark Office: "Trademark/Service Mark Application, Principal Register, With Declaration" and "Amendment to Allege Use Under 37 CFR 2.76, With Declaration." These are relatively short (two pages, including instructions).

The filing fee currently in effect is $175.00 for each class of goods or services for which application is made.

Information can be obtained from, and applications sent to:

U.S. Department of Commerce
Patent and Trademark Office
Washington, DC 20559
ph (202) 707-9100

Registering on Basis of Intention to Use

In late 1989, the Trademark Law Revision Act of 1988 went into effect. One of the most significant changes over prior law is the provision for allowing trademark registration based on a bona fide *intent to use.* Formerly, the trademark had to have a history of actual use in commerce before it could be registered. Now, the Patent and Trademark Office will process the application for intended use the same as it does for an existing trademark application. The mark will be published to allow for comment or opposition from interested parties. If competitors offer no opposition or if such oppositions are disallowed, an notice of acceptance wil be issued. The applicant will then have six months to file a verified statement that the mark is being used in commerce. This statement must specify the date of first use in commerce, the goods or services to which the mark is being applied, and the mode or manner of use (e.g., on packages or in printed ads). If the statement is approved by the Patent and Trademark office, registration will be approved. After use begins, the applicant must submit, along with with the form previously described, specimens evidencing use and a fee of $100.00 for each class of goods or services in the application.

The term of the Federal trademark registration is 10 years with renewals for 10 years. However, between the fifth and sixth year after the date of registration, the registrant must file an affidavit stating the mark is currently in use in commerce. If no affidavit is filed, the registration will be cancelled.

If a trademark is registered but is never used or is used and then abandoned, the owner may lose the right to it, depending on the circumstances. Federal law provides that, if a registered trademark has not been used for two years, the nonuse is prima facie evidence of abandonment (Geren 1984B).

Statutory Grounds for Rejection

Your application for trademark registration may be refused if the mark or term applied for:

1. Does not function as a trademark to identify the goods or services as coming from a particular source; for esample, if the matter applied for is merely ornamentation.
2. Is immoral, deceptive, or scandalous.
3. May disparage or falsely suggest a connection with persons, institutions, beliefs or national symbols, or bring them into contempt or disrepute.
4. Consists of or simulates the flag or coat of arms or other insignia of the United States, or a state or municipality, or any foreign nation.
5. Is the name, portrait, or signature of a particular living individual, unless he has given written consent; or is the name, signature, or portrait of a deceased president of the United States during the life of his widow,

unless she has given her consent.

6. Resembles a mark already registered in the Patent and Trademark Office to such a degree that it is likely, when it is used on or in connection with the goods of the applicant, to cause confusion, or to cause mistakes, or to deceive.

7. Is merely descriptive or deceptively misdescriptive of the goods or services bearing the trademark.

8. Is primarily geographically descriptive or deceptively misdescriptive of the goods or services of the application.

9. Is primarily merely a surname.

A mark will not be refused registration on the grounds listed in numbers 7, 8, and 9 if the applicant can show that the mark has become distinctive through use in commerce so it now identifies to the public the applicant's goods or services.

A mark refused registration on points 1, 7, 8, and 9 might find a place in the *Supplemental Register*, which contains terms or designs considered capable of distinguishing the owner's goods or services, but that do not yet do so. But, a term or design cannot be considered for registration in the *Supplemental Register* unless it is in use in commerce in relation to all the goods or services identified in the application, and an acceptable allegation of use has been submitted. If a mark has been registered on the *Supplemental Register*, the registrant may bring suit for trademark infringement in the Federal courts, or may use the registration as a basis for filing in some foreign countries, but none of the other benefits of Federal registration apply.

Color as a Trademark

In recent years, the concept of color *alone* as a trademark has received expanded recognition and validation by the courts. Take as an example the pink color of the fiberglass insulation popularized by Owens-Corning. According to Bullwinkel (1992), "To claim trademark rights in a color, you must meet a three-part test: (1) there must not be a competitive need to use a particular color in the marketplace; (2) the use of a particular color must be nonfunctional and non-utilitarian; and (3) you must show that the color mark has in fact achieved secondary meaning."

Legal decisions vary. In 1991, a U.S. Circuit Court ruled that NutraSweet couldn't use trademark law to protect the blue color of its single-serving packet, and the Supreme Court declined to review the decision. The manufacturer of Prestone antifreeze lost an attempt to claim the yellow color of their plastic jugs as a trademark. Procter & Gamble did not prevail in an attempt to prevent rivals from using a pink-colored remedy for stomach disorders. At the time this is being written, a case is in preparation that will allow a court to determine whether or not blue plastic grips on tools can be protected as a trademark.

Could color be used a trademark in a food product? Although this concept seems a bit far-fetched at first consideration, there does seem to be a possibility that color could be made a legally accepted trademark for certain types of foodstuffs. Blue potato chips, red popcorn, or purple bread are items that are certainly feasible from a technical standpoint and, indeed, some of them have been offered in limited markets. The economic value of such a trademark might be essentially nil because of the presumed low level of consumer acceptance of such nonconventional-appearing products, but other variations might have more appeal. It is obvious that white potato chips, white popcorn, and white bread could never be claimed as color-trademarked products because of widespread prior (and current) use by hundreds of manufacturers.

The previous discussion refers to single colors not arranged in a particular pattern, i.e., covering the whole surface of a product or otherwise not forming an image. It appears there is general agreement that combinations of colors, especially in some sort of recognizable pattern, are definitely acceptable as copyrights. Also, a single color used as part of a unique design can be protected as part of the trademark, e.g., McDonald's "golden" arches.

Identifying Trademarks Used on Commercial Products

To prevent trademarks from becoming generic terms, leading to loss of trademark protection, it is necessary to identify the trademark as such wherever it appears. A satisfactory notice is achieved by using the R in a circle or the superscript TM in small capitals. These denote, respectively, Registered and applied-for trademarks. Apparently, attaching the circled R to unregistered marks could result in civil liability, although cases where penalties are applied for this kind of activity must be very rare. The trademark notice should appear at least once in each piece of printed or other visual matter, and preferably at the place where the trademark appears for the first time. The notice should be positioned immediately adjacent to the trademark; for example, position the circled R on the shoulder of, or adjacent to, the last letter of the registered mark. Also, it is acceptable to use the phrase "Registered U.S. Patent and Trademark Office" or the abbreviation "Reg. U.S. Pat. & TM Office." For marks as yet unregistered, the ^{TM} should be positioned similarly.

Many corporations or even unincorporated firms assume that their company names, as registered with the appropriate state agency, can be freely used under any circumstances and in connection with any product or service of the corporation. This is not necessarily the case. The name can't be used if it infringes the existing rights of other parties, and there are other limitations as well.

COPYRIGHTS

Perhaps your company has developed a piece of art, a description, an instruction booklet, an advertisement, or other piece of reproducible graphics material to promote a new product. If this material has some value as a sales aid, it may be possible to use copyrighting as a means of preventing competitors from imitating it, provided the intellectual product is original and meets certain other requirements. The item need not have any "artistic" or esthetic value by any standard whatsoever in order to be copyrightable. Copyrights have the great advantage that they have much longer terms than patents and, in practice, than most trademarks. Also, unlike trademarks, a period of prior use is not necessary to make them effective and continuation of use is not necessary to keep them in force.

The Copyright Office of the U.S. publishes a number of circulars that cover most of the important aspects of copyright protection. The text in these publications is, generally, clear, succinct, and relatively free of legal jargon. Anyone contemplating filing an application for copyright protection should have these circulars on hand for ready reference. Copies are (as of the time this is being written) free. Write or call: Register of Copyrights, Library of Congress, Washington, DC 20559. Some of the circulars you may find to be helpful are:

Copyright Basics (Circular R1)
New Classification System for Copyright Registrations (Circular R1c)
New Copyright Registration Procedures (Circular R1d)
Publications on Copyright (Circular R2)
Copyright Fees (Circular R4)
Obtaining Copies of Copyright Office Records and Deposits (Circular R6)
The Effective Date of Registration (Circular R8a)
Duration of Copyright Under the New Law (Circular R15a)
Renewal of Copyright (Circular R15)
Extension of Copyright Terms (Circular R15t)
Highlights of the New Copyright Law (Circular R99)
How to Investigate the Copyright Status of a Work (Circular R22)
Highlights of U.S Adherence to the Berne Convention (Circular 93)

Much of the following discussion will be based on information found in the above listed publications, but the reader should be aware that changes invalidating the statements made below could occur at any time. Always verify the current validity of any regulatory provision before making a final decision upon the actions you wish to take.

The term of protection lasts for the balance of the author's life plus 50 years after his death. If the work has been made "for hire" (or if it is pseudonymous or anonymous), protection is granted for 75 years from the date of first publication or 100 years from the date of creation, whichever is less. There are some exceptions.

If the work is of a copyrightable nature, legal protection is assumed to exist as soon as the work is created. It is not necessary to register the work to obtain a certain degree of copyright protection. There are a number of advantages to registering the work, however. For example, suit for infringement cannot be brought until the work has been registered; also, statutory damages and attorneys' fees cannot be recovered in an infringement action unless the work was registered within three months of publication.

Copyrightable works are classified as follows:

Class TX. Nondramatic literary works. Most sales aids developed for promoting new products would fall in this very broad category. The class consists of all types of published and unpublished works written in words or other verbal or numerical symbols. Examples are fiction, nonfiction articles, poetry, periodicals, textbooks, reference works, directories, catalogs, advertising copy, and compilations of information.

Class PA. This category includes published and unpublished works prepared for the purpose of being performed (as on a stage or before a television camera). It includes musical works, dramatic works, dances, motion pictures, etc. Scripts and other preliminary intellectual work leading to commercials advertising new products might fall into this class.

Class VA. Works of the visual arts are included in this class. Published and unpublished pictorial, graphic, and sculptural works are examples. Within this class are pictorial or graphic labels and advertisements, as well as works of artistic craftsmanship. If the product has some utility as a tool or piece of furniture, it might be copyrightable provided it has pictorial, graphic, or sculptural features that can be identified separately from and capable of existing independently of the utilitarian characteristics of the article. If you have produced a shovel with a sculptured handle, it may be possible to copyright the handle as a sculpture but not the shovel qua shovel. It seems obvious that any number of copies (e.g., cast replicas of a sculpture) may be made under copyright protection. Remember that the copyright office is not supposed to act as an art critic in deciding whether or not an item is worthy of this protection.

Class SR. This category of sound recordings is used for registrations of published and unpublished works in two situations: (1) where the copyright claim is limited to the sound recording itself, and (2) where the same copyright claimant is seeking to register not only the sound recording but also the musical, dramatic, or literary work embodied in the sound recording. With one exception, "sound recordings" are works that result from the fixation of a series of musical, spoken, or other sounds. The exception is for the audio portions of audiovisual works, such as a motion picture soundtrack or an audio cassette accompanying a filmstrip; these are considered an integral part of the audiovisual work as a whole and are registrable in class PA. It seems reasonable to assume that advertising jingles for radio or television would fit in this class.

Class RE. Renewal registrations. In some cases, there is an opportunity to renew a copyright that is in the process of expiring. If this option is available, the method of obtaining the extension is very simple and the cost is nominal.

In general, the registering of a copyright is extremely easy and cheap, much different from the case with patents. Taking as an example Class TX works, registration consists of submitting two copies of the work along with a completed form and a small fee to the department at the Library of Congress which handles these documents. The application is a two page form requiring answers to about 15 questions, the latter being answerable by most authors in a few minutes or by a team of lawyers in a few days. Granting the application appears to be a matter of routine with the government examiner merely determining whether or not the form is filled out correctly, but it still takes several weeks under normal circumstances. Of course, the copyright can be attacked as invalid by anyone who believes, or pretends to believe, that it infringes on a prior work or consists of material which is in the public domain.

Although legal assistance is not really required if the work is of a standard nature — book, story, poem, etc. — and poses no other problems, many persons will want to consult a legal adviser to insure that they are obtaining the maximum protection and have not invalidated their protection by failing to follow some obscure provision in the copyright law.

The "new" copyright law (Copyright Revision Act of 1976) has teeth in it. A copyright owner has several remedies for infringement. These include monetary recovery in the form of either actual or statutory damages, injunctive relief, and the impounding and destruction of infringing articles. The infringer may also be subject to criminal penalties for certain acts. The copyright owner can bar the importation of infringing articles. If a foreign firm duplicates a machine you have developed but not patented, you may not be able to prevent importation of the machine but you may be able to bar the entry of a published instruction manual or maintenance manual which duplicates the one you have developed for the machine, provided you have copyrighted your manual.

Among the more novel uses of copyrights that have been attempted is the copyrighting of internal corporate records in order to prevent them from being widely disseminated. In some recent product liability cases, Japanese manufacturers have copyrighted testing and safety documents to prevent plaintiffs from publicizing this information or using it in additional suits. It is not clear whether court decisions will sustain this strategy. The usual position is that mere data, or obvious or conventional arrangements of data (as arrangements of names alphabetically in a telephone book or of numerical data in a simple and obvious tabular form) cannot be copyrighted.

The dividing line between patents and copyrights, formerly very clear, has been blurred in recent years by proposals, actions, and court decisions relating to copyright protection for computer chip designs and for the masks used in making these chips. These materials would seem to be strictly utilitarian (rather than "artistic") in nature, and as such not protectable under current patent and copyright laws. Obviously, the much greater simplicity of applying for, and the much longer protection obtainable under, copyright law makes it greatly preferable to the patenting process, when a choice is available.

PATENTS

There are three very different kinds of patents: for living plants, for ornamental designs, and for industrial inventions.

Even though plant patents are related to the very interesting and important field of genetic modification, which has a number of economically important implications for growers and processors of food vegetation, their legal position and scientific nuances are too complex to cover in this book. There has also been a fairly recent decision by the Patent and Trademark Office to grant patents on new forms of animal life produced by genetic engineering or other human intervention. The first such patent was granted to Harvard University for a highly cancer-prone mouse, an obvious boon to environmentalists.

Design patents cover ornamental inventions used to improve, or at least change, the outward appearance of products, machines, etc., without improving the utilitarian qualities of the item. Since design patents may have some value for developers of new food products, they will be discussed briefly before beginning the more extensive section on product and process patents.

Design Patents

Design patents are granted to persons who have invented new, original, and ornamental designs for an article of manufacture. This type of patent protects only the appearance of an article, not its structure or utilitarian features. One might get a design patent on a hoe handle carved in the shape of Marilyn Monroe, but if the function or intent of this innovation is to improve the gripping firmness or the strength of the handle (rather than its pure esthetic value), a design patent is not the appropriate vehicle to use in obtaining patent protection. The proceedings leading to the granting of design patents are about the same as those relating to other patents, but there are a few differences. Similarities and differences will be discussed in the following paragraphs.

The drawing of a design patent must conform to the same rules as other patent drawings, except that no reference characters are required. The specification is short and usually follows a set format. Only one claim is permitted and it, like a process patent claim, must conform to a standard format. Apparently, there is no requirement that the product confer any special advantage or improvement over existing items, except for the esthetic advantage perceived or imagined by the inventor.

Although there are a number of essential elements in the application for a design patent, the nub or crux of the submittal is the drawing. According to Burge (1984). the drawing(s) must include a sufficient number of views to provide a complete disclosure of the entire appearance of the article. "Nothing regarding the shape, configuration, and surface ornamentation of the article may be left to conjecture." The whole part of the basic object should be included in the drawing, but dotted or broken lines may be used to indicate those parts of the object that are not to be covered by the desired patent, with a statement in the text (or drawing caption) declaring the function or meaning of the dotted or broken lines.

A design patent is effective for a period of 14 years, and no fees are necessary to keep it in force, contrary to the situation with regard to industrial patents. Of course, there is a fee for the initial issuance of the patent.

Relatively few design patents are granted for food products, but the contours of a puffed corn snack, the shape of a pretzel, or the pattern on the outside of an ice cream cone might be the subject of a design patent if only the visual attributes of the usual prior form of the product are changed.

Insofar as the type of matter protected is concerned, design patents are actually closer to copyrights than to product and process patents. For example, it is certainly possible to copyright an original work of art, such as a sculpture of a horse, but it would seem to be possible, also, to obtain a design patent on this object. If the horse is intended for use as, e.g., a doorstop, a paperweight, or a bookend, then the design would certainly be patentable; if it exists only as a work of art, however, there might be a question about patentability. If it has improved functionality as a paperweight, for example, because it has the new shape, its acceptability as the subject of a design patent would be doubtful.

Both the patent and the copyright would protect the innovator against unlicensed copying of the object, but a copyright is much easier and cheaper to obtain and the length of protection is much longer. The reader may logically object to the apparent implication that the "contours of a puffed corn snack, the shape of a pretzel, or the pattern on the outside of an ice cream cone" could be the object of a copyright, but on the other hand, why not? Stranger things have been copyrighted and the copyright upheld in court.

Burge (1984) points out, however, "A great many designs are patentable, but are not copyrightable, because these designs do not meet the copyright requirement of separate identification and independent existence." and "Copyright protection for the design of a useful object is obtainable to the extent that the design incorporates pictorial, graphic, or structural features that can be identified separately from, and are capable of existing independently from the utilitarian aspects of the article."

One company has recently applied for a copyright on "new and original and artistic three-dimensional works which consist of decoration on the outside surface of its [Sweet Street Desserts, Inc.] pies and the unique interior structure of such pies." The concept of using a copyright (rather than a design patent or perhaps trademark law) to protect the appearance of a food product is evidently a new approach and it is not yet known whether the copyright will be granted and whether it can be defended successfully in court if it should be granted (Holleran 1993). It can be predicted that the copyright office will be flooded with applications from food companies if Sweet Street does prevail in its attempt. Copyrights are generally easier, quicker, and cheaper to obtain than design patents, and have a much longer life.

Process and Product Patents

In legal language, the category of patentable subject matter encompasses "any new and useful process, machine, manufacture, or composition of matter, or any new and useful improvement thereof."

The patent type of greatest interest to most readers of this book will be utility patents. They give to the patentee (or his assignee) the exclusive right to practice the invented process or make the invented product for 17 years from the time the patent is granted (some exceptions are made due to special circumstances). The operative word is "exclusive;" in a sense, a patent confers a negative right – it can be used to prevent others from making, using, or selling the patented invention but it does not give an absolute right to make or market the invention.

If, for some reason, the invention conflicts with an existing law or regulation, or requires the incorporation of some process or product patented by, or otherwise being the property of, another person or firm, the patent will not override these limitations and the inventor may not be able to legally market his new invention even though his patent is perfectly valid. This might be the case when an inventor develops an improved part or component for an item that has been patented by someone else. Once the older (original) patent expires, the inventor of the part may be able to make the whole item and sell it, or the owner of the original patent may license the use of the improved part during the remaining life of his patent. Conversely, the new inventor may obtain a license to produce the original item, put his new part in it, and sell it as an improved version. These situations crop up in the computer business rather frequently.

Utility Patent Applications

Utility patent applications, to be sent to the Commissioner of Patents and Trademarks, include as a minimum: (1) A written document comprised of an oath or declaration and a specification (description and claims), (2) A drawing, in those cases in which a drawing is necessary, and (3) the filing fee. The specification and oath or declaration must be legibly written or printed in permanent ink on one side of the Paper. The applications as received are numbered in sequence and the applicant is sent a filing receipt that includes the serial number and filing date.

The oath or declaration must affirm that the inventor believes that he or she is the original and first inventor of the subject matter of the application. If it takes the form of an oath, it must be sworn to by the inventor before a notary public or other officer authorized to administer oaths. A declaration (which need not be notarized) may be used instead of an oath as part of the original application (1) for a patent involving designs, plants, and other patentable inventions, (2) for reissue patents, (3) when claiming matter originally shown or described but not originally claimed or (4) when filing a divisional or continuing application.

Blank forms for applications are *not* supplied by the Patent and Trademark Office.

Who is the inventor?. It is said that only the inventor(s) may apply for a patent on an invention – but there are exceptions. If the inventor dies, an application may usually be filed by his legal representative. If the inventor is insane or otherwise incompetent, his guardian may file. If an inventor refuses to participate in the filing process, or if he cannot be found, a joint inventor or a person having a proprietary interest (as an employer) in the invention may apply on behalf of the unavailable inventor. Unless one or more of these relatively unusual circumstances applies, the actual inventor(s) must apply for the patent.

If two or more persons participate in discovering the invention, they must apply as joint inventors. A person who only makes a financial contribution is neither an inventor nor a joint inventor, though he may own part or all of a patent as soon as it is granted as a result of assignments made at some time during or before the process by the actual inventors.

The inventor's employer may have some right of ownership in a patent depending upon the employee's contract of employment and other factors, but the employer is not the inventor merely because of the contract. Is the inventor's right-hand man (or woman) a co-inventor if this subordinate did all the experiments, machined the parts, tested the finished machine, etc.? Maybe. The inventor is the person who contributes the essential elements of patentability, namely originality of idea and reduction to practice. Sorting out who did what, when the identification of an inventor becomes a matter of controversy, is a matter of fact to be elucidated in a court of law, and may take years of litigation before it is resolved.

Specification. The title of the invention should be as short and specific as possible and should appear as a heading on the first page of the specification.

The specification (description and claims) must include a written description of the invention and of the manner and process of making and using it. The specification must be in such full, clear, concise, and exact terms as to enable any person skilled in the art to make and use the invention.

The specification must describe in complete detail a specific embodiment of the process, machine, manufacture, composition of matter, or improvement invented, and it must explain the mode of operation or principle whenever applicable. The best mode contemplated by the inventor of carrying out his invention must be set forth. For example, a range of temperature conditions within which the invention will operate may be set forth, but if there is a variation of effectiveness within this range (as is usually the case), the optimum temperature must be disclosed.

It is the author's opinion that many inventors, or their attorneys, needlessly risk the status of their invention by making the ranges of operable conditions unduly wide. If it is stated that the invention is operable within the range of, say, 100°F and 200°F when there is no record of successful experiments being conducted at either end of this range, and it can be shown that the invention does not, in fact, operate at 100°F, doubt is cast on other points in the disclosure and there may be some question as to whether the invention exists. These discrepancies may not be picked up by the examiner — they probably won't be — but, in the event you are giving a deposition to an opposing attorney in a patent case, you can be certain you will be questioned about the method used in establishing the ranges and optimums stated in the patent application.

An abstract of the technical disclosure in the specification must be set forth in a separate paragraph under the heading, "Abstract of the Disclosure." This must be placed on a separate page immediately following the claims.

A brief summary of the invention as claimed, indicating its nature and substance, should precede the detailed description. This may include a statement of the object of the invention.

The specification will end with one or more claims particularly pointing out and distinctly claiming the subject matter that the applicant regards as the invention. These claims are to be brief descriptions of the subject matter of the invention, without unnecessary details yet reciting all essential features necessary to distinguish the invention from what is old. Claims are the operative part of the patent. Novelty and patentability are judged by the claims, and questions of infringement are judged by the courts on the basis of the claims.

Filing fees. The filing fee, which must accompany the application, consists of a basic fee and additional fees. The basic fee allows the presenting of twenty claims, including not more than three in independent form. An additional fee is required for each claim in independent form in excess of three, and an additional fee is required for each claim in excess of twenty. Multiple dependent claims also mandate additional fees.

For the purpose of calculating fees, (1) a claim is in singularly dependent form if it incorporates by reference a single preceding claim which may be either a dependent or an independent claim, and (2) a multiple dependent claims or any claims depending from a multiple dependent claim shall be considered as separate dependent claims in accordance with the number of claims to which reference is made.

If the owner of the invention is a "small entity," the filing fees are reduced by half provided the owner files a verified statement justifying this status. A small entity is described as either (1) an independent inventor, (2) a small business concern, or (3) a non-profit organization.

Even for a small entity, the fees are far from being negligible. At the time this is being written, the basic filing fee is $690, reduced to $345 for small entities, and there is every indication that substantial increases will be made in the near future. There are additional fees for every conceivable change or addition to the application, and maintenance fees of hundreds or thousands of dollars must be paid after 3.5, 7.5, and 11.5 years or the patent will lapse (a grace period of 6 months following the due date is allowed). Once the patent expires, anyone can practice the art that was disclosed.

Priority and timeliness. The patent application must be filed within one year from the time the invention is first publicized or commercialized. What constitutes first publication or first public use often becomes a difficult legal problem and is, in any case, too complex to discuss here.

The patent systems of many foreign countries are based on the "first to file"principle, i.e., the first person to file an application gets the patent even though he may have reduced the invention to practice after some other person. There have been attempts to change the U.S. system to this method of establishing priority, but so far the efforts have been unsuccessful.

If two or more patent applications are filed for the same invention, a proceeding known as an "interference" comes into play. In most cases, an interference results when an application is found to describe an invention very similar to one disclosed on an application that is still in the examination stage, i.e., the earlier one has not yet been granted. Only about 1% of all patent applications filed lead to an interference.

Under certain circumstances, an interference procedure can be instituted when a patent has already been granted, provided the earlier patent has not been granted more than a year before the filing of the conflicting application. Each party to an interference proceeding must submit evidence showing when the invention was made. When no evidence is submitted, the earliest date that will apply to the invention will be the date the respective application was filed. Priority is determined by a board of three interference examiners on the basis of evidence which has been submitted. The losing party may appeal to the Court of Customs and Patent Appeals, or file a civil action against the winning party in the appropriate U.S. district court.

Some Pros and Cons of Patenting

Patent law is very complex, and the composing of a patent application that will secure the maximum amount of protection without committing actionable perjury is a highly specialized profession. Lawyers' fees and patent office fees can be onerous burdens for the individual inventor. As a general rule, filing for a patent should not be undertaken unless the inventor is willing to commit a considerable amount of time and money to the process.

To secure a valid patent, the inventor must disclose all conditions necessary for the patent to operate successfully. If some critical detail has been omitted, so that a knowledgeable reader of the patent cannot duplicate the product or process from the disclosed description, the patent can be voided on the basis of inadequate disclosure. Thus, the inventor or his assignee gives away all of his secrets in return for the questionable grant of right of sole use for seventeen years. If the patent is voided because of some technicality or declared to be invalid for any reason, the inventor is left with a worthless piece of paper and a stack of legal bills, for which he has exchanged trade secrets of some value – perhaps of great value. For this reason, each inventor must carefully consider whether it is preferable to file for a patent or to treat the invention as a trade secret. This decision must, of course, be made within a year from the time the invention is reduced to practice, after which the point becomes moot because a patent will not be granted..

A few patents turn out to be very valuable, but most patents yield nothing, or almost nothing, to their inventors. Part of the problem is that infringers, if there are any, must be located and, if necessary, taken to court to get them to pay the royalties the inventor thinks he deserves. The defendant will probably claim the patent is invalid, for any one of a number of possible reasons, or may admit the patent is valid but claim he did not infringe it. In the past, courts have been only too willing to go along with claims of non-infringement or invalidity. According to Mangels (1986), in the 1950s and 1960s patents were judged to be valid in only about 30 to 40% of cases decided in appellate courts and, in some cases where the courts found the patents to be valid, they determined that the patents had not been infringed, so that the plaintiffs prevailed only in about 30% of the cases decided by the appellate courts. The situation has changed somewhat in recent years, and the percentage of cases in which the defendant has been found to have infringed a valid patent is considerably larger. Part of this change is due to the establishment in 1983 of a new federal appellate court to hear all of the patent appeals from all of the district courts throughout the country. This court has tended to rationalize and make more uniform the state of patent law.

Many scientists and technologists regard patents as important and valuable additions to their resume. If an obliging employer is paying the bills and providing legal assistance, the inventor's effort involved in obtaining patents is probably well repaid by the resulting addition to his or her vita. If the inventor must bear the expense of prosecuting a patent application, however, there is a real question as to whether a patent is a suitable investment. Of course, for many independent workers, who doubtless have no hope of getting their results published in refereed journals, the gaining of one or two patents may provide useful prestige points for improving either social or professional status. The independent inventor must, however, consider the possibility that the application will not be granted, resulting in a complete loss of the time and money spent in prosecuting the application.

Do It Yourself?

It is nearly always difficult, costly, and time-consuming task to prosecute a patent application. A patent attorney or agent is essential, or almost so. It should be pointed out that patent agents who are not attorneys may represent applicants before the Patent and Trademark Office if they have been approved and registered by that Office. Perhaps some very simple invention (if there are any) might be patented without the aid of an attorney or agent, but the author does not recommend considering this course even if such conditions apply. This means that the absolute minimum cost for supporting a patent application from filing through granting (or final rejection) will amount to a few thousand dollars, and most will run much more.

Time is another problem. A few years ago, it was customary for even straightforward patents with a half dozen or so claims to take 3 or 4 years to finalization, but now some are being granted in about 18 months, a significant improvement. Even so, a large amount of your valuable time is going to be required for filling out the necessary papers, responding to queries, doing additional experimental work to fill in the gaps pointed out by the patent examiner, etc.

If you decide to manage your own application, you may wish to refer to "Patent It Yourself," one of the most complete discussions of the process (Pressman 1988). Be sure you are using the most recent edition – the copy you find in your local library may be out of date.

Sources of Patent Information

The principal and most current source of original information on new patents is the Patent Gazette, which contains abstracts of all issued patents. Reviewing this bulky periodical to find the abstracts of patents related to some specific project can be difficult and time-consuming, partially because the classification system used is inefficient and out of date. The abstracts published in the Gazette also fail, in many cases, to adequately summarize the contents of the patents. As a result, it may be more effective, though certainly more expensive, to use other databases or consultants to find patents relevant to your specific problems. Some of these sources are (taken in part from Anon. 1992):

1. Chemical Abstracts Service, P. O. Box 3012, Columbus, OH 43210 (ph 800-848-6538)
2. Derwent, Inc., Suite 401, 6845 Elm st., McLean, VA 22101 (ph 703-790-0400)
3. DIALOG, 3460 Hillview ave., Palo Alto, CA 94304 (ph 800-3DIALOG)
4. Pergamin ORBIT InfoLine, Inc., 8000 Westpark dr., McLean, VA 22102 (ph 800-421-7229)

A good summary of current patent regulations can be found in the booklet, "General Information Concerning Patents." This booklet, which is periodically revised, can be obtained for a few dollars from the Superintendent of Documents, Washington, DC 20402, or possibly from the nearest government bookstore.

TRADE SECRETS

There is no federal statute covering or even defining trade secrets, but legal actions can taken under the common law and under some state statutes for a violation of your trade secrets rights. Under the existing situation a trade secret can be almost anything used in the course of conducting business – a process, formula, product, technique, customer list – that gives a company a competitive advantage, is kept secret within the company, and is not generally known in the industry (Perham 1977).

For example, a company that has developed a completely new process or product or has made improvements in an old process or product has certain property rights in those developments even though nothing about them is patentable. If some part of the discovery appears to be patentable, management must decide whether to file for a patent or to treat the invention as a trade secret. If a patent application is filed, every detail necessary to allow another person to duplicate the invention must be disclosed. If the discovery is to be treated as a trade secret, every effort must be made to keep as many details as possible from becoming known outside that small group within the company who must know about the discovery in order to practice it for the benefit of the company. There are few exceptions to the general guideline that the law views all information as being freely available to the public and not susceptible of protection.

A company that does not use every reasonable effort to keep its trade secrets confidential is likely to find the courts unsympathetic to its problem when a competitor introduces a duplicate process or product. The real problem with relying on trade secrets, is that a second inventor who independently discovers the product or process and promptly files a patent application may be able to obtain a valid patent on the invention in spite of your apparent priority of practice. According to Geren (1984A), the deciding factor (whether to file or not) may be the life cycle of the product or process – if the expected life cycle is three years or longer, a patent is probably preferable to the trade secret option. Life cycle, in this instance, means the time before the object of the trade secret becomes obsolete or is superseded by advances.

Although it was formerly considered that issuance of a patent released any trade secret rights the inventor had in the subject of the patent, the results of recent cases indicate that it may be possible to retain some trade secrets in the patented process, particularly in specific conditions which improve upon the ranges disclosed in the patent specifications. If the disclosure included a specification of 200° to 300°F as an acceptable processing range, but extensive pilot plant studies showed that 295°F was really the only efficient processing temperature, there is at least a possibility the optimum condition could be considered a trade secret. If the improvements are developed after the patent has been filed, the inventor is in a somewhat better position. Readers who wish to learn more about the current thinking on this point are advised to consult a patent attorney.

One step in establishing the company's intention to maintain certain information, processes, etc., as trade secrets, is to inform each employee of the importance of keeping confidential the art he will learn and practice within the company. According to Bloom (1993), the company should adopt ". . . comprehensive policies covering access to and use of confidential information . . . These policies need to be updated continuously to reflect changes in technology."

Pooley (1982), says that a company must be able to describe what it claims to be a trade secret because, "Without a precise description, you cannot prepare an appropriate plan to protect it, or expect others to respect your rights . . . The courts will not recognize trade secret claims that are not Sufficiently specific." He recommends having a carefully developed and monitored program of protecting proprietary information, continually seeking to define your trade secrets, and acting as though your trade secrets are valuable.

One list of measures that a company should take to protect trade secrets is as follows:

(1) Establish a written trade secret security program.

(2) Learn work histories of technical applicants so that a charge of using a prior employer's trade secrets will not be brought.

(3) Require that employees sign nondisclosure agreements; the document should inform them of the kinds of information they are not to disclose.

(4) Inform all personnel of the firm's trade secret security program.

(5) Maintain registers of trade secrets and lists of persons who have access to them.

(6) Limit distribution of information, and more specifically, access to trade secrets, to those people whose jobs involve use of the secret processes, data, formulas, etc.

(7) Use "confidential" stamps on documents, drawings, etc., that are expected to be treated as trade secrets. Simply stamping "confidential" on every memo, plan, ad, etc., may, however, lead a court to question whether or not you know what trade secrets you do have.

(8) Keep your trade secret files locked when they are not in use. Restrict access to "need to know" persons.

(9) Post and restrict access to trade secret areas such as laboratories, production facilities, etc.

(10) Review proposed outside presentation of technical papers for possible unauthorized trade secret disclosures. Warn presenters of the danger of "off-the-cuff" comments on these subjects.

(11) Require outside suppliers, laboratories, and sub-contractors to agree to use all reasonable efforts to maintain confidentiality with respect to your trade secrets.

(12) Record the location and possessor of trade secret documents and destroy these papers when they are no longer required.

(13) Review trade secret agreements with employees who are departing the company for any reason (including retirement) and notify their future employers of their trade secret obligations.

Perhaps it will be worthwhile, perhaps only amusing, to quote the distinguished author R. Austin Freeman (in "The Funeral Pyre"), ". . . the people who have trade secrets are those who make a profound mystery of simple processes that any schoolboy could carry out with one showing. That is the necessity for the secrecy."

FOREIGN PROTECTION

The protection your innovation or trademark can receive outside the U.S. varies widely from country to country. There are treaties in effect with many nations that can allegedly penalize and restrain illegal competition more or less in accord with U.S. principles. In a number of cases, the countries have enacted laws closely resembling (except for the language, of course) regulations earlier promulgated in the U.S. Obtaining timely action and recompense for damage is another matter. Greenberg stated (1993), "One government estimate reportedly says that misappropriated trademarks, patents and other unfair competition is a $100-billion problem each year. The type of protection offered by the laws of other countries varies, as does the usefulness of resorting to foreign courts for timely remedies."

According to Geren (1982), "The first consideration for an inventor is not whether one may file, but whether or not the foreign right would have some value." He also points out that in many countries, a sale anywhere in the world of the goods or disclosure anywhere in the world of the process prior to the filing date to which the foreign application is entitled bars the obtaining of a valid patent in that country. This is called the absolute novelty approach and is effective in nearly all European countries. The term of patents vary from country to country.

It is very clear that you need to employ an attorney having experience in foreign patent procedures if you intend to pursue this method of protecting your invention.

TRADE DRESS

There is an amorphous concept that could be called unfair-competition-by-copying-trade-dress that is sometimes claimed by plaintiffs who feel their general or overall approach to some market has been copied by a competitor. It is a last resort in cases where trademarks, copyrights, trade secrets, or patents cannot be shown to have been infringed. As an example, consider a greeting card manufacturer who believes that a competitor has copied a line of their cards and used promotional techniques, market positioning, sales strategies, and other selling techniques very similar to their own. No copyright infringement can be shown, trade secrets are not applicable since only publicly known matters have been copied, there are no patents, and the trademarks are not similar, but the general appearance, demographics of customers, type of sales outlets, and spirit of the original cards are thought to have been deliberately copied. It is in such situations that unfair competition based on intentional copying of trade dress can be alleged as a basis for claiming damages.

In 1990, a restaurateur in lower Manhattan sued a nearby restaurant that was, he claimed, using a menu, product composition, company name, and other operational details similar in many respects to those he had originally developed and was currently using.. The principal products of the two companies were giant-sized pancakes in many different flavors. The originator of the concept evidently proved his case and reached a settlement that caused management of the competitive restaurant to make significant changes in its operation. News reports indicated that the copying was, in this instance, extremely flagrant, including use of exact duplicates of formulas, printed menus, etc.

A Supreme Court decision in June of 1992 appears to have significantly expanded the protection which will be given for trade dress. The court ruled in favor of Taco Cabana International, which had claimed that Two Pesos Inc. had copied the Taco Cabana menu and some design features of its restaurants. It is difficult to understand how the individual elements of the Taco Cabana operation could be regarded as having any uniqueness, so the decision must have been based on a perception that the total concept of the Taco Cabana restaurant, the particular combination of all the many obvious and previously-used elements of design, operation, menu, etc., was copied. The court's rationale, as described in news reports, was that a company seeking to prevent imitations of its appearance needed to prove only that its appearance is "inherently distinctive," not that consumers viewed the appearance or mode of operation as a unique mark of that company. Taco Cabana had previously taken a competing chain of restaurants to court because, it was said, the competitor imitated Taco Cabana's "festive eating atmosphere."

Zaharoff (1993) gave some clarification on the Taco Cabana case. He said, the Supreme Court held that, when product or service trade dress is inherently distinctive (rather than, for example, commonplace or functional), the vendor claiming rights in that dress need not prove that consumers already recognize that the trade dress is associated with a single vendor. That is, the court distinguished between inherently distinctive trade dress and trade dress with secondary meaning. A mark acquires secondary meaning when an appreciable number of consumers understand it is being used as a trademark, rather than in its primary descriptive sense.

It may seem to be farfetched that the concept of trade dress could be a possible method for protecting product innovation, yet, who would have believed five years ago that practices or things not shielded from imitations by trademarking, or patenting, or copyrighting, or trade secret privileges would be given a high degree of protection by the highest court in the land.

LICENSING DISCOVERIES

The "renting" of the right to use intellectual property such as patents and copyrights is generally called licensing and is a thriving branch of commerce. There are two diametrically opposite approaches to the licensing of discoveries in new product development. Innvovations made by another company's research staff can be licensed for use by your company, or discoveries made within your company can be offered for use by other firms if your management feels they do not have the resources to fully exploit the potential new product.

Such opportunities arise when a new process or product is developed (or partially developed) before the company's management realizes that it cannot fully exploit the development's potential. If this happens very often, it is clearly a sign of poor R&D direction, but an infrequent occurrence can be the result of a change in management goals, a flash of genius, etc. The problem then becomes one of obtaining some return for the expended time and money, or, putting it more positively, of maximizing the return on a valuable discovery.

Licensing a discovery to another firm better suited to utilization of the development's advantages is a possible course of action. Few manufacturers are set up to merchandise ideas. There are, however, consulting firms that specialize in the dissemination of information on such opportunities to persons or manufacturing concerns who may be interested in them. Generally, the consultant will also handle the preliminary negotiations. Fees are likely to be rather high and are not always contingent upon successful completion of licensing or sale.

Among the attorneys who have published discussions of licensing strategies and procedures is Greenstein (1978), who listed the following reasons for licensing:

(1) A proprietary product is being manufactured and sold, but full exploitation of the innovation cannot be realized by the owner because of limited marketing capability.

(2) The technology lies outside the scope of your company's principal business interests, but is compatible with the activities of another company.

(3) Commercial development of the innovation would require expenditures beyond your company's financial resources.

(4) Cross-licensing in which two companies exchange one technology for a another more suited to their firm's operations. This sometimes occurs as a way to settle litigation with minimal cash passing and little or no damage to the balance sheet.

(5) Your company developed proprietary methods for manufacturing an item important to its own specifications (as an essential ingredient not commercially available), but cannot be bothered with the actual manufacture. A supplier may be licensed to produce the item(s) in accordance with the desired specifications.

Other firms will also offer to license their discoveries to your firm. Their representatives seldom, if ever, make their first approach to the R&D department, preferring to negotiate with the CEO if possible, or, failing that, the head of Marketing. Marketing departments and general management often do not realize that ideas obtained from other companies in this way may require substantial amounts of additional development work before they can be fitted into their organization's production and distribution patterns. Much of the work already done may be completely inapplicable to your company's needs.

Any such offers coming into your company should be evaluated by every department that would normally review an in-house R&D proposal before a decision is made to proceed with negotiations. A thorough investigation of the technical feasibility must be made — do NOT rely on the consultant's claims that all production problems have already been solved. If you are allowed to do so, make direct contact with the selling company's R&D people who handled this project and find out the true reason for selling the idea. If you are not allowed to do so, consider the denial as an important warning signal. Does the offering party or company have actual physical samples of the product, if it is a product? If samples are not available, find out the reason. Is the original developer backing out because present or future regulatory problems are perceived?

If your firm has a policy against licensing discoveries made by outsiders, no one in the firm (including the CEO) should ever agree to examine or evaluate ideas submitted to them by consultants, outside inventors, or the like. If the firm should subsequently introduce or try to patent a product resembling in the slightest degree the prototype that has been submitted or demonstrated or explained by the outsider or his agent, a claim for royalties, damages, and other compensation is almost sure to result.

When it is necessary, for some reason, to get further details about an innovation that has been exposed to public scrutiny, (for example, a new product idea has been described in the newspapers), it is best to delegate the investigation to a third party, such as the firm's lawyers. The representatives should be instructed to get signed clearances from all parties contacted by them and to keep confidential all matters which they do not have clearance to reveal. Many cases could be cited as examples of the need for caution, see Deogun (1993) for a recent one.

BIBLIOGRAPHY

ANON. 1963. Will your notebook stand up in court? Chem. Eng. Progr. *59*, No. 7, 12-13.

ANON. 1979. Highlights of the new copyright law. Circ. *R99*. Copyright Office, Library of Congress, Washington, DC 20559.

ANON. 1984. Copyright basics. Circ. *R1*. Copyright Office, Library of Congress, Washington, DC 20559.

ANON. 1989. Basic Facts about Trademarks. U.S. Dept Commerce.

ANON. 1990. Guide for Patent Draftsmen. Patent Office, U.S. Department of Commerce.

ANON. 1991A. More lawyers caution: watch that trademark. Wall Street Journal, Feb. 8, 1991, p. B1.

ANON. 1991B. Product liability suit can't be brought against publisher. Wall Street Journal, July 16, 1991, p. B5.

ANON. 1992. Understanding chemical patents. Today's Chemist at Work *1*, No. 3, 24-27.

ANON. 1993. Americans earn rising share of U.S. patents. Chem. Eng. News *71*, No. 8, 16.

BARBER, H. L. 1989. Copyrights, Patents and Trademarks Worldwide. TAB Books, Blue Ridge Summit, PA

BLOOM, C. A. 1993. Do you want to know a trade secret? Informationweek *447*, p. 84.

BRESNICK, S. R. 1987. Patents and the food industry. Food Technol. *41*, No. 4, 118-121.

BULLWINKEL, G. 1992. Trade dress protection expanded by new decision. Packaging Digest *29*, No. 11, 20, 22.

BURGE, D. A. 1984. Patent and Trademark Tactics and Practice, Second Ed. John Wiley & Sons, NYC

DEOGUN, N. 1993. Small firms sue, charging theft of ideas. Wall Street J. *92*, No. 39, B-1, B-6.

DREHKOFF, W. D. 1980. The importance of patent literature. Cereal Foods World *25*, No. 3, 94-95.

DUNNIGAN, J. A. 1988. Bringing your product to life. I. Patenting your product. Entrepreneur *17*, No. 3, 58, 60-62.

EVANS, L. E., Jr. 1986. Patents, Trademarks, and Copyrights. Gunn, Lee, and Jackson, Houston, TX

FERGUSON, T. W. 1991. Liberating inventors or shackling progress with paperwork. Wall Street J., March 5, 1991, p. A15.

FIELD, T. G., Jr. 1989. Avoiding Patent, Trademark and Copyright Problems. U.S. Small Business Admin., Office of Business Development.

GEREN, G. S. 1982. Foreign patent protection — an overview. *24*, No. 6, 29.

GEREN, G. S. 1983A. Barring imports by using patents, trademarks, and copyrights. Ind. Res. Dev. *25*, No. 11, 31.

GEREN, G. S. 1983B. The copyright/high tech interface. Ind. Res. Dev. *25*, No. 4, 29.

GEREN, G. S. 1984A. In-house processes — are they secret? Res. Dev. *26*, No. 11, 27.

GEREN, G. S. 1984B. Problems in trademark enforcement and abandonment. Res. Dev. *26*, No. 9, 31.

GEREN, G. S. 1984C. Proprietary rights in start-up companies. Res. Dev. *26*, No. 12, 29.

GEREN, G. S. 1984D. Searches — what are they? Res. Dev. *26*, No. 2, 33.

GEREN, G. S. 1984E. The patent-trade secret interface. Res. Dev. *29*, No. 5, 31.

GEYELIN, M., and HAGEDORN, A. 1989. Trademark law change is taking effect. Wall Street J., Nov. 15, 1989, p. B8.

GILLIS, K. F. 1993. Kosher foods find a broader market. Food Business *6*, No. 4, 18.

GREENBERG, E. F. 1993. When does competition cross the line? Packaging Digest *30*, No. 5, 22, 24.

GREENSTEIN, H. 1978. Licensing new product technology. Industrial Res. Develop. *20*, No. 6, 122-127.

HOLLERAN, J. 1993. Suit questions design rights. Food Business *6*, No. 13, 15.

JOHNSON, R., and POUND, E. T. 1992. Hot on the trail of trade-secret thieves. Wall Street J. *90*, No. 31, B1-B2.

LAMBERT, W. 1990. Investing in patents to file suits is curbed. Wall Street Journal, May 30, 1990, p. B4.

LEVINE, R. E. 1989A. Hey, can I tell you this in confidence. Packaging Digest *26*, No. 7, 28, 30.

LEVINE, R. E. 1989B. Patent infringement can be a costly mistake. Packaging Digest *26*, No. 4, 31-32.

LEVINE, R. E. 1989C. Trademark law has changed significantly. Packaging Digest *26*, No. 11, 30.

MANBECK, H. F., Jr. 1992. General Information Concerning Patents. U.S. Dept. Commerce.

MANGELS, A. J. 1986. The changing world of patents. Food Technol. *40*, No.7, 72-74.

MARSHALL, E. 1991. The patent game: raising the ante. Science *253*, 20-24.

MAYNARD, J. T., and PETERS, H. M. 1991. Understanding Chemical Patents. American Chemical Society, Washington, DC

PERHAM, J. 1977. How to protect corporate secrets. Dun's Review *110*, No. 2, 49-50, 76-77.

PETROSKI, H. 1992. The evolution of artifacts. Am. Scientist *80*, No. 5, 416-420.

POOLEY, J. 1982. Trade Secrets: How to Protect your Ideas and Assets. McGraw-Hill, NYC

PRESSMAN, D. 1988. Patent It Yourself, Second Ed. Nolo Press, Berkeley, CA

PRESTON, E. 1982. A Writer's Guide to Copyright. The Writer, Inc., Boston, MA

SCHLESINGER, A., Jr. 1989. The judges of history rule. Wall Street J., Oct. 26, 1989, p. A16.

SHAVER, A. M. 1977. The new copyright law – some pertinent questions. Plastics Eng. *33*, No. 4, 16.

SELZ, M. 1990. Somebody made off with the pancakes that took TriBeCa. Wall Street J., July 11, 1990, p. A1.

SHAVER, A. M. 1977. The new copyright law – some pertinent questions. Plastics Eng. *33*, No. 4, 16.

SITTIG, M., NOYES, R., and STECHER, P. 1980. Corporate patent confrontation and litigation. Cereal Foods World *25*, No. 3, 96-97.

WADE, W. 1965. Industrial Espionage and Mis-use of Trade Secrets, 2nd Edition. Advance House Publishers, Ardmore, PA

WOO, J. 1993. Trademark law protects colors, court rules. Wall Street J., Feb. 25, 1993, p. B1, B10.

ZAHAROFF, H. G. 1993. Secondary meaning of some trademarks. Wall Street J., 6 June 1993, page A16.

PRESENT AND FUTURE TRENDS

INTRODUCTION

There is value in analyzing the course of past trends in customer reception of new products and applying the knowledge so gained to the prediction of future trends, with the view of designing new products that can be introduced early in the preliminary stage of food fads and fashions. This chapter is intended to give the reader a a brief introduction to the various strategies that might be of value to developers of new products.

Novelty itself is an attractive feature. Many retail customers will, if the mood is right and the purse is full, buy a product simply because it is something they have not tried before – of course, it must be identifiable as being part of a class they are familiar with and it must not be so bizarre as to cause uneasiness. Manufacturers using this type of appeal as their main selling point for a new product should realize that novelty is self-limiting – once a new food has been tried, it is not novel anymore for that consumer. Pointing out the unusual features of a product is a useful strategy for getting the first sale, then, if the product is more desirable than competitive items for other reasons, follow-up sales will likely follow. In this connection, it is worthwhile to consider the trend of sales in bread, a product that might be expected to have relatively few opportunities for innovation.

A study presented at the 1993 meeting of the W. E. Long Independent Bakers' Cooperative (Anon. 1993A), stated that U.S. bread sales (in dollars) during 1992 showed the following changes in the categories named: heat-and-serve +35.4%, Italian/ French/sourdough breads +5.4%, flavored/sweet breads +5.3%, pocket breads (e.g., pita) +4.1%, vegetable/ fruit/nut breads +3.5%, rye/pumpernickel -1%, low-calorie -2.3%, grain/ protein/fiber -5.2%, and egg breads/challah -17.6%. Among the interesting changes seen in this survey are the current downward moves in some types of bread that have been the subject of intense publicity in the not-too-distant past.

In the last few years, new product introductions in the U.S. have mostly emphasized health benefits and nutritional claims (not always the same thing) and high-priced "gourmet" variations, often with considerable market success. Ethnic variations of familiar foods often appeal to the curiosity of consumers, and all large food manufacturers have made some attempts to establish or follow major trends in this category. Prestige and gourmet foods (the two categories overlap and are often indistinguishable) are trendy by their very nature, and offer particularly attractive opportunities to manufacturers because the buyers of such products are not as price-sensitive as the average consumer.

HEALTHY FOODS

Health claims are faddish in nature – sales of products the appeal of which is based on some particular health benefit fluctuate according to the enthusiasms of the moment. News media managers, unscrupulous scientists and physicians, and other opportunists are eager to use for their own advantage the gullibility and lack of specialized knowledge of those people whose major concerns are health status and vitality. The supply of dietary superstitions and hypotheses is unlimited. As the audience's attention flags, new or re-cycled ideas are introduced, and, if the reception is promising, are exploited to the fullest extent.

It will often be observed that different factions having ready access to the news media will put forth contradictory views at the same time. Individual consumers differ in their susceptibility to various pressures and in their resistance to changing their minds when faced by confusing situations of this type, so the mythical "average" consumer must be relied upon when discussing overall trends in healthy foods.

Cholesterol and saturated fats seemed to be the chief "unhealthy" food components a year or two ago, but are becoming less important as the focus of publicity shifts to other quasi-nutritional subjects. Supermarket sales of fresh eggs (an egregious villain in the high cholesterol game) rose 15% during a 12 week period during 1993, leading an industry expert to say "We're seeing consumer attitudes shift, consumers are beginning to be less frightened of food in general and more confused about what is good for them" (Deveny 1993A).

Increasing the fiber content of foods has been another popular approach to health improvement through nutritional modification. High fiber content can be achieved with large additions of grain by-products such as wheat, oat, or corn bran, or by adding more exotic ingredients like psyllium seed husks, purified cellulose (from wood, sugar beet pulp, etc.), and pectinous materials. Products containing increased calcium or decreased sugar have also been successfully introduced into the marketplace. Other entries in the health-with-happiness field include low cholesterol, reduced sodium, and low fat.

Future developments will follow new dietary fads as they become entrenched in the popular imagination. The exact direction of these changes cannot be predicted with a high degree of confidence, but products that have their caloric content reduced or that are useful for some other reason in a dietary regimen intended to reduce body weight will be popular as long as there is an abundance of good food for all consumers.

The product development technologist would be well-advised to give nutritional science discoveries second place to trends in media emphasis when attempting to find a basis for predicting the direction of changes in consumer demands. These currents in news coverage are related to (1) a follow-the-leader mentality among writers for the popular press and (2) the activities of public relations firms that have a financial interest in either the discrediting of or the popularizing of a particular kind of food.

Attention should be given to the declared intentions of competitors, if only to use them as negative indicators. In the 1993 Prepared Foods R&D Investment Survey (Anon. 1993B), it was reported that the percent of respondents projecting R&D-spending increases of 20% or more over the 1994-1995 period for healthy food development of the type indicated were: 52% reduced calorie, reduced fat, or diet foods; 26% reduced cholesterol foods; 23% reduced sodium foods; 21% all-natural (no additive) foods; 20% removing additives from existing products; 17% removing saturated fats from existing products; 14% removing preservatives from existing products; 13% added fiber foods; 7% reformulating trans-hydrogenated fats from existing products; 6% removing tropical oils from existing products; and 5% organic foods. The reader should note that these figures apply to the percentage of companies in the panel expecting to take action on foods in the named category, not to their projected percentage increase in R&D spending on these projects.

GOURMET AND PRESTIGE FOODS

The categories of gourmet and prestige foods offer opportunities for trend-following in which directional changes in the trends are often slower to develop and thus easier to predict and exploit than fads in healthy foods.

There are times when it is difficult to differentiate between gourmet foods and prestige foods. Generally speaking, a prepared food qualifying for the definition "gourmet" would be expected to meet higher standards than ordinary foods in ingredients, preparation methods (more careful measuring, more skillful forming, etc.), freshness (except where aging is a necessity), and presentation (decoration, etc.). Gourmet foods will, almost always, be higher in cost (often, much more expensive) than ordinary foods of the same type.

Prestige foods are those the choice of which confers upon the consumer (at least in his own mind) the cachet of keener aesthetic discernment (better taste), more informed selection (more intelligent), better access to limited supplies (can afford to pay more), or other characteristics indicating personal superiority of some type (membership in some elite group, greater public recognition, etc.). Such foods are by their nature and as part of their consumer attraction higher in price than ordinary foods of the same general type, and their consumption is almost always publicized in some manner, if only to one's dinner guests.

Gourmet cookies give us a well-defined example of this type of food. These items result from formula variations that give highly enriched cookies, i.e, they contain relatively large percentages of chocolate, fruit preserves, cream cheese, butter, and the like. They will continue to sell in good volume regardless of their health connotations. It is to be expected that these products will probe the most advanced frontiers of richness and elaborate design, so that the boundaries between candy bars, ice cream confections, French pastries, "Danish," and cookies will be even more blurred than it is today.

Prestige cookies, as represented by varieties that have had their appearance improved by decorative enhancements or that contain perishable components such as whipped cream and genuine buttercream fillings may attract additional consumer interest. There should be an increase in gourmet and prestige cookies that can be distributed only in frozen form because of their limited shelf stability, since this form of distribution allows the commercialization of many exotic and expensive types that would otherwise be impractical.

The boutique cookie shops that started to appear in shopping malls and other high traffic locations during the 1970's provide an opportunity for supplying consumers with freshly baked cookies having characteristics different from anything that can be achieved in shelf-stable products, though at a much higher price; this market seems to be substantially static at present. It is possible, though rather unlikely, that other types of foods might be merchandized in the same manner. The features the manufacturer would want to emphasize are freshly prepared, better quality, unique varieties, and easy availability.

ETHNIC SPECIALTIES

In this category are included foods that are regarded, rightly or wrongly, as being typical of some group having what is thought to be readily recognizable cultural characteristics. The food will usually be promoted as being typical of a nation or other region (Russia, Hungaria, India, Japan, etc.) or of a religion (Hindus, Moslems, Mormons, Jews, etc.). Sometimes, the ethnic designation is considered to apply to groupings of humanity based on other criteria (cowboys, farmers, hunters, etc.).

Among the prominent examples in this category is bagels. They have a distinctly ethnic appeal. Apparently, they are regarded as being a primarily Jewish-oriented food, though the product itself probably originated in central Europe among populations that did not have a particularly high ratio of Jews. The current ethnicity attribution seems to have originated in New York City during the late 1800s, a time and place where there was a closely-knit culture of middle-European Jews. Whatever the background, there is a current trend to expand the bagel market geographically and to proliferate varieties. Typifying this tendency is the recent introduction of "Bagelettes," described as an all-purpose snack combining the properties of breadsticks, bagels, and crackers (Anon. 1992); this item comes in the varieties sesame, Parmesan and Romano, cajun, garlic, onion, and rye.

Caution should be the watchword in following this trend. The unadorned bagel has relatively little to recommend it, all examples being tough, low in flavor, and rather uninspiring in appearance as compared to most other varieties of yeast-raised rolls, and their characteristic shape and size limits their application to combination foods (such as sandwiches; it is difficult to put a frankfurter on a bagel). Although their resemblance in shape to doughnuts may suggest to some development technologists that a similar wide range of line extensions would be successful, the innate characteristics of bagels seem to work against such diversification.

Currently, tortillas are in a strong upward trend as their distribution and consumption expands far beyond the original Hispanic markets. This trend will probably continue until the country has been fully exposed to the food in all of its various disguises, then a stabilization and possibly a mild decline in total consumption volume will occur. Tortillas can not successfully compete in flavor with good quality yeast-leavened breads, but their shape, especially their adaptation to all sorts of "wrapping-around" combinations, gives them an advantage that is almost unique. Pita breads and other flatbreads can, however, compete to some extent in these applications. The masa (corn) tortilla will continue its decline relative to consumption of wheat flour tortillas.

FOODS MEETING PHILOSOPHICAL OR RELIGIOUS TENETS

Although it might seem that trends in those food varieties conforming to special philosophical and religious requirements would be slow to develop – that consumers responsive to such claims would be very conservative and difficult to reach– there are indications that occasional opportunities do occur for the development technologist to take advantage of changing situations in these areas.

Sometimes an unfilled need exists or could be created because members of a particular group do not clearly differentiate between what is acceptable and what is not, while others may wish to become even more "pious" (in either a religious or philosophical sense) than their fellow members and therefore demand specially formulated and prepared foods that carry to extremes the publicly enunciated tenets of their organization. Or, leaders of the group may see fit to issue pronouncements of new dietary laws, or new interpretations of existing practices, that call for specially formulated foods or beverages.

Examples of food preferences based on philosophical or religious concepts are the demand for foods free of animal materials, as required by vegetarians, and the requirement for foods that do not combine dairy ingredients and meats, as specified by strictly kosher diets. Other limitations include the avoidance of alcohol (which may extend to flavor extracts used in the formulas) by Muslims and Mormons, and the avoidance of beef by Hindus.

In some cases, foods currently being produced may meet the requirements of a group that the manufacturer wishes to target as a specific new market. Only a small change in the wording on the label may be required –

"contains no animal-derived ingredients," "free of dairy ingredients," or "contains no beef [or pork]" may serve the purpose. In other cases, the buyers may require more assurance, so the label may have to include phrases such as "certified kosher by Rabbi Moishe Birnbaum" or the trademarked kosher designations K and circled U. At least one organization now offers trademarked certification of foods meeting Muslim dietary laws.

It is often found, however, that formulation modifications and processing changes will be needed to meet the special food needs of the targeted religious or philosophical market. These requirements can become very complicated and may involve abstruse points that do not seem logical to the scientifically trained product developer. Recourse to published material may provide inadequate, out-of-date, or otherwise faulty guidance. As a means of overcoming these problems, the manager of new product development may find it desirable to hire a consultant having status and recognition among members of the organization. The consultant would not only have the responsibility of determining the correctness of formulas and processes (based on his group's standards) but would also provide assistance in obtaining distribution and publicity for the products among members of the flock and might be able to suggest new products.

Postum, the famous coffee substitute composed principally of roasted grain and molasses, was formulated specifically for a religious group that had adopted the practice of avoiding coffee. Of course, it ultimately reached a much wider market, including people who wished to avoid caffeine because of health concerns.

Many times, obtaining an entrée into these markets is impossible because of the existence of special non-publicized arrangements that exist between leaders in the group and current suppliers.

DIETETIC FOODS

Dietetic foods vary in their application from "healthy" foods in that the former are designed for the use of persons having a disease or physiological disorder demanding the use of specially-formulated foods or the avoidance of certain common foods. Diabetes is a typical example of a diseased condition requiring a regulated diet, in this case necessitating the absence of, or a very low level of, sugars.

There are many diseased conditions in which the symptoms can be ameliorated or prevented by the use of special foods or, more correctly, by the avoidance of certain food constituents. As the average American becomes older, more overweight, more debilitated by drugs and alcohol, broken down by work or exercise, under greater stress from other causes, and is subjected to more diagnostic tests and unneeded medications, it is certain that the demand for dietetic foods will increase. The question is, which of the categories of dietetic foods will become the best candidates for new product development. A full discussion of every possible dietetic food is certainly not possible in this small volume, but a list of some of the types of foods fitting into the broader category will be given in the following paragraphs as a basis for further research by the reader, if there is an interest in this category.

It should be pointed out at the beginning of this discussion that some dietetic foods, such as low residue diets for surgical patients, require specialized marketing staffs and entail a higher risk of legal liability than regular foods. There is also the possibility that the foods, as marketed, could be regarded as "new drugs" leading to astronomical costs and long delays for obtaining regulatory approval. For this reason, a dietetic product should always be positioned as a "food" having special nutritional properties; no claims should be made that it is a treatment for a disease.

Among those types of products that the food developer may wish to consider when trends in dietetic foods are being evaluated, are:

(1) Hypoallergenic foods. These are items purporting to have decreased allergenic properties or to be substitutable for another food having allergenic properties. An example would be an imitation chocolate cookie having no cacao-derived ingredients and therefore suitable for persons who are allergic to chocolate.

(2) Infant foods. There are special labeling requirements for infant formulas and other foods designed to meet the special nutritional needs of young children. These requirements are significantly more onerous than those for regular foods, but opportunities may exist for the manufacturer wishing to follow developing trends in market demand that may result from newly-published findings in infant nutrition.

(3) Foods intended for use in modifying caloric intake or body weight. There will always be a demand for foods to make you skinny or to allow you to gain weight more easily, the trick is to foresee the direction fads in dieting will take and be prepared to adapt to them without delay. Books on new ideas in dieting, if they receive good publicity, will almost always lead to marketing opportunities, but they may be temporary in nature, lasting hardly long enough to get a food of the proper type into the stores. It is important for the manufacturer not to be caught with merchandise designed to meet an outmoded fashion, because such material has essentially no marketability.

(4) Diabetic foods. There are many versions of foods designed for use by diabetics. This market has been exploited for so long, and the scientific understanding of the needs of diabetics has been static for so many years, that the trend in introductions is basically flat. Opportunities for new products generally involve the use of known artificial sweeteners in the production of low-sugar or no-sugar versions of conventional foods and beverages.

Diabetic versions of nearly all common foods have been developed and are widely distributed, so the new product manager who wants to enter this field is confined to replacing some of the more exotic foods or to offering foods that are of higher quality (in terms of taste, texture, and appearance) than existing diabetic products.

(5) Low- or no-sodium foods; low salt or no added salt. Labeling requirements establish many of the parameters for these foods in considerable detail. Direction of new product trends would seem to be restricted largely toward adaptations to take advantage of new nutritional information or to exploit in a line of products the value of newly discovered seasonings that duplicate the flavor of salt without having the undesirable physiological effects of sodium.

(6) Reduced fat or low fat foods. The market for these products overlaps to a considerable extent the target population for foods intended to be used in a dietary regimen for reducing body weight. But, there are differences. Physicians will sometimes specify a diet low in "triglycerides," or emphasize the need for avoiding saturated fats or trans-fatty acids. Obviously, the best way to achieve all of these goals is to rely on a diet much lower in total fat than the average. This is difficult to accomplish without the use of highly specialized (and usually unappetizing) foods. The trends in low-fat and non-fat foods are at present rather hard to identify and describe, but appear to be mostly in the direction of using viscosity-modifying hydrocolloids to replace fats. In some products, particularly in salad dressings, very satisfactory organoleptic results have been obtained. Following the trend would consist in developing formulas or processes that apply to other kinds of foods or that give organoleptically improved (better taste, texture, or appearance) versions of competitive products.

(7) Gluten-free products. This is an example of a category of foods expected to appeal to a very limited market, in this case, persons having fundamental physiological problems related to the ingestion of "gluten." Celiac (coeliac) disease is a serious disorder triggered by wheat, rye, durum, or oat flour, and perhaps by other substances. It is not an acquired disease, but appears to be present from birth, and it can be fatal. The disease is not a allergy, though it appears similar in some respects to allergies.

There is no discernible trend of introductions of gluten-free foods in the marketplace. There do appear to be potentials for good quality simulations of bread and cake that do not contain wheat flour or other substances triggering the symptoms of coeliac disease, but the potential consumers will be hard to reach and the customers will be widely dispersed.

(8) No cholesterol. Trends in this category are complicated and irregular because of the fluctuating nature of the scientific evidence regarding the effect of cholesterol on health — particularly, respecting its role in the development of atherosclerosis. The diffuse and imprecise nature of the definition of "cholesterol" and the problems encountered in analyzing for it are also troubling.

It is simple enough to make many kinds of foods that are "cholesterol-free." This requires only the elimination of any animal fats or ingredients derived from animal fats. Non-usable materials would, of course, include such materials as egg yolks as well as all meats. Most authorities agree that cholesterol is not found in any plant product, including vegetable oil, or at least is present in negligible amounts. Some published analyses have, however, shown traces of cholesterol or cholesterol-like substances in vegetable lipids. Actually, a pure animal fat (pure in the sense that it contains only triglycerides) wouldn't contain cholesterol, but lard, butter, and other shortenings made from animal fats will include some of this substance as an impurity originating in the tissue used as a raw material for the fat.

The formulation changes required in switching from an animal shortening to a partially hydrogenated vegetable oil (for example) are, in most cases, minimal. The developer must exercise great care in screening minor ingredients and processing aids if a "no cholesterol" claim is going to be made. Some emulsifiers are made from animal fats and thus may contain traces of cholesterol. In bread baking, pan greases may contain animal fats, and washes used to impart glossiness to the crust may include materials containing some cholesterol.

CHANGES IN DIMENSIONS

Making the piece size larger has been one way of differentiating products from the competition, and along this line we have seen larger cookies (approaching cakes in diameter) and larger muffins. This direction of product modification seems to have become somewhat less popular as ever larger proportions of the general population become more interested in limiting their caloric intake. There may still be opportunities in special niches where the size increase has functions other than giving the consumer more to eat in one piece.

The opposite trend, that of downsizing product piece size, has been in effect for 4 or 5 years now. The reduction in dimensions has been applied not only to individual pieces but also to the weight or volume of the retail container's contents. If we look at cookies, we can find many kinds, including the ubiquitous sandwich cookie, in piece sizes much smaller than the traditional variety. Thus, cookies of the Oreo type, the size of which has been standardized for decades, now appear in diameters nearer that of a quarter-dollar than of a silver dollar. Downsizing of this particular item no doubt makes it easier for small children to consume, and may also appeal to parents who feel less guilty giving a smaller portion of an "indulgence" product to their children.

Because of differences in proportion of components (base cake and creme) and the non-linearly-related thickness adjustment necessary to give satisfactory consumer acceptance, the weight relationship of components in the small and large cookies is by no means proportional to the ratio of the respective areas that could be calculated from the diameters of the traditional and the midget pieces.

Miniaturizing candy bars, such as Snickers, to give a "snack size," sold as multiples of ten or more dump-packed in plastic bags, has achieved only limited success, according to Deveny (1993B). She says, "Supermarket shoppers avoided nearly every major brand of the small candy bars during the back-to-school season [fall of 1993]." This downward trend may, however, be attributable to other reasons than the downsizing; there is a general impression that purchases of candy for Halloween trick-or-treat distribution fell in 1993, and the sales of small candy bars depend to a great extent on this seasonal surge. Also, there appears to be a general trend away from candy because of nutritional concerns. For whatever reason, the trend to smaller sized versions of candy bars may have peaked.

Offering smaller retail packages, such as pouches containing a couple of ounces of cookies, caters to the same market segment that has made portion packs of potato chips, popcorn, corn chips, and the like so popular.

SWITCHING CATEGORIES

A product having only limited success in one marketing category, such as snack foods, may gain additional volume by being packaged for and marketed to customers currently interested in another category, such as health foods. Changes in formulas and processing methods (shape, size, etc.) may be needed to successfully compete in the new category. Since recent trends in consumer preferences seem to lead away from indulgence foods and toward health foods, there have been many examples of snack food manufacturers trying to re-position their existing products (or some minor variations of them) as health foods – natural, organic, enriched, low-fat etc.

A study by Hwang (1993) describes the results of attempts by manufacturers to switch snack crackers (or a modification of them) into the category of salty snacks (potato chips, pretzels, corn curls, tortilla chips, etc.), and vice versa. This reporter quotes an industry representative, "There's a growing recognition that both snack crackers and salty snacks deliver many of the same benefits to consumers. . . It doesn't take a very huge step to move [crackers] closer to the [potato chip] category." One result of this recognition has been a proliferation of new chips intended to appeal to the consumer of crackers, and, conversely, new crackers with many of the characteristics of chips. Among these new entries (as listed by Hwang) are Nabisco's Mr. Phipps Pretzel Chips and Tater Crisps, SunChips from Frito-Lay, Munch'ems and Wheatables from Keebler, and Borden's Graingers.

COMBINATIONS HAVING CONVENIENCE ADVANTAGES

Among the representatives of this category recently entering the market are: (1) Cheese spreads in combination packs with pretzels, bread sticks, crackers, and the like, intended to be eaten by scooping out the cheese spread with the cereal item; (2) Cans of tuna in a package also containing crackers and pouches of mayonnaise, pickle relish, and other condiments; and (3) Portion-size cups of yoghurt with a packet of dry cereal attached.

Combinations of this sort do seem to provide added conventience, especially for the lunch box contingent, and should be explored in more detail if the products made by your company have characteristics that fit them for such positioning. New products not specifically designed for combination packets may have properties that suit them for this market as well as for the originally targeted positions.

Placing dissimilar products in the same package may create very serious problems with both shelf-life and distribution methods that are not immediately obvious. These problems include, but are not limited to, interchange of moisture and flavor between the components, restriction of shelf-life to that of the most rapidly deteriorating component, and the possible need for storage at refrigerator or freezer temperatures of relatively large masses of material that would otherwise be distributed, stored, and displayed at ambient temperatures.

PERILS IN ADVANCING BEYOND THE BORDER

It is tempting to take advantage of a developing trend by going one or two steps beyond the currently popular version of some fad. If the populace is emptying the shelves of three-grain bread, it may seem obvious that it is desirable to introduce a four-grain or five-grain bread. This is one of the easiest ways of pre-empting a developing pattern and, indeed, it sometimes turns out to be successful. It is often a much easier and usually a less risky alternative than entering the market with a totally new concept, and it is nearly always much less expensive than a full-scale product development project that requires the conducting of marketing tests intended to determine the potential of a truly innovative food. But, there comes a point when it is intuitively pointless to carry a trend a step further. If there is a ten-grain bread on the market, will the average consumer (or any consumer) really believe that an eleven-grain bread is more desirable than the previous choice? Is an additional type of grain more likely to influence the consumer's decision than the color of the package, the price, a convenient location on the store shelf, or any of the many other elements affecting his choice?

The developer may ask, "But what is the risk? The 11-grain bread will sell at least as well as the 10-grain bread, so at, even under the worst case scenario, no market share will be lost by moving a step further along the trendline." Even though the new product may involve only minor modifications in an existing product, such changes will always result in the incurring of some risk and some cost.

The charges for designing new packaging, the problems involved in obtaining a new ingredient (probably one having limited availability), the unknown effects of the new ingredient on processing equipment and on the food's texture, flavor, and appearance, and the regulatory consequences of the food modification are all factors that will have to be considered as possible influences on costs and risks when even a relatively minor change is made. The assumption that no adverse effect on consumer acceptance will result from a product incorporating a small change in the direction of an existing trend is not sufficient reason for making the change; some real advantage in sales volume or profit must be reasonably anticipated.

There will always be a reliable though limited market for specialty foods that look terrible and taste worse but are "home-made" or are claimed to be "healthful," "environmentally-friendly," "natural," or "organic" and these market niches will continue to be exploited with food products that can make combinations of such claims. No doubt we will be hearing more about foods that are politically correct, gender sensitive, or contain recycled ingredients. Such marketing niches are based on fads – the fads may last 10 or 20 years, but they are fads nonetheless. Manufacturers who intend to exploit such niches must constantly survey the marketplace so they can quickly withdraw from areas no longer popular, otherwise they risk being caught with large amounts of merchandise that can never be sold, not even in the "Everything for $1" stores.

Deveny (1994) points out the temporary and uncertain nature of the market for celebrity-recommended foods, e.g., Paul Newman's salad dressings, Sinatra's spaghetti sauce, and Lee Iacocca olive oil. This is not a subject that would ordinarily be of great interest to food technologists, although they might receive an order from the CEO or marketing head to formulate a product suitable for selling under the name of a currently notorious person. The basic approach would probably be to use a standard formulation, one known to have wide appeal – trying to design a truly superior product would fall under the heading of casting pearls before swine. Sales of these products are likely to have only a fleeting popularity, if any, and warehouse stocks should be kept to a minimum and frequently monitored.

BIBLIOGRAPHY

ANON. 1992. Bagelettes: an all-purpose snack. Food Business *5*, No. 16, 27.

ANON. 1993A. Analysis shows bread growth by category. Bakery Production and Marketing *28*, No. 5, 140.

ANON. 1993B. Nutrition priorities. Prepared Foods *162*, No. 2, 49.

ANON. 1993C. The pretzel's time has come. Baking Buyer *Jan. 1993*, p. 14-15.

DEVENY, K. 1993A. Americans take a fresh crack at eggs. Wall Street J. *92*, No. 74, B-10.

DEVENY, K. 1993B. Chocolate bar sales fell sharply in quarter. Wall Street J. *92*, No. 121, B-5.

DEVENY, K. 1994, Many celebrity foods enjoy only 15 minutes of fame. Wall Street J.*93*, No. 12, B-1.

FULLER, G. W. 1993. Ingredients and green labels. Food Technology *47*, No. 8, 68-71.

HARWOOD, J. 1991. Bakers serve up a variety of new products. Farmline *12*, No. 7, 8-10.

HWANG, S. L. 1993. Hybrid snacks try to be tasty, healthy. Wall Street J. *92*, No. 25, B-1.

KUHN, M. E. 1993. Developing products from the outside in. Food Business *6*, No. 20, 44.

NEFF, J. 1993. Line extensions: enough already? Food Business *6*, No. 13, 20-21, 24.

OTTO, A. 1993. Spilling the beans in Spokane. *6*, No. 16, 35.

SLOAN, A. E. 1993. The changing dynamics of healthy eating. Food Technology *47*, No. 11, 42.

TWISS, B.C. 1986. Managing Technological Innovation, Third Ed. Longman Group Ltd., Harlow, Essex, England.

chapter fifteen

THE TEN COMMANDMENTS OF PRODUCT DEVELOPMENT

The ten commandments have been expanded to a baker's dozen to permit the inclusion of more of the important factors that the technologist must keep in mind during product development work.

(1) Keep complete records of experiments. Jot down details (especially of quantities, temperatures, times) as the experiment is being conducted. One person of the team that is conducting an experiment should be designated as the principal record keeper or record collator even though several people may be recording data at any given time.

(2) Identify all samples from the very beginning of their production and maintain this identification (recorded in a lab notebook) until the sample has been destroyed. An item without a clear, unambiguous identification is not a sample – either destroy it or describe it in a manner such that its problems are obvious to anyone who examines it. Never put a sample identification number on a container that has a dubious provenance.

(3) Identify and fully describe the ingredients used in an experiment. The description should include the age and storage history of the ingredient, among other data.

(4) Recognize that the package (including necessary label statements) is as much a part of the finished product as the ingredients you put in the mixer. This should keep you aware that package development cannot be separated from food development.

(5) Put dates on all your records, including casual notes. Don't write anything on scrap paper unless you are absolutely sure you will be discarding it immediately.

(6) Recognize that shelf-life includes the entire period from the time the product emerges from the processing line until it is eaten by the consumer. Recognize also, however, that engineering a product to be stable under extreme conditions or for very long periods of time will often result in an item that is unacceptable to most consumers from day one.

(7) Understand that a product requiring any form of preparation by consumers will surely encounter extremely diverse conditions during preparation and that some failures of the product to perform as expected are inevitable. It is not humanly possible to forsee all the things a consumer can do wrong. As a first approach, never assume that all purchasers will understand the directions, no matter how simply they are expressed. Any set of directions consisting of more than 3 or 4 simple declarative sentences of five words each will confuse the majority of consumers.

(8) Recognize that cost is always a consideration but that some of the greatest successes in food marketing have been achieved with upscale pro-

ducts aimed at categories that have undergone a continual cheapening over a period of many years (bread, ice cream, etc.).

(9) Accept the fact that many consumers simply are not capable of recognizing fine nuances of flavor and texture, and, if they were, wouldn't pay a premium for them.

(10). Never, never, place a toxic material (certain chemical reagents, for example) in anything even remotely resembling a food product unless that item is kept under close control at all times. Casual visitors to the lab, or intruders, are inclined to taste anything left lying around if it looks as though it might be a food product.

(11) Develop new products compatible with existing production lines if management insists, but don't forget the consumer doesn't care how much unused capacity your company has, and a product that doesn't fit consumers' needs will not be successful no matter how happy it makes your accounting v.p. or production manager. The starting point for a new product should be (ideally) a determination of the requirements of the consumer, not an assessment of the limitations of existing production facilities.

(12) All test panels (focus groups, taste test groups, expert panels, etc.) are biased and non-representative of the "average" consumer by the very fact that they cannot, as a practical matter, consist of every potential consumer of your product testing the product under the conditions existing when they would normally consume it. This does not mean that such panels should not be used; it means that the results obtained should be viewed as indicative rather than definitive. Statistical analyses extrapolating the results of such tests so as to predict the results of the performance of a product in the marketplace should be taken with a grain of salt.

(13) When you have irrefutable evidence that a proposed product is either unacceptable to the targeted consumer group or impossible of development in the time or under the conditions existing in your company, do not hesitate to pull the plug on the product, no matter how much time and money has been spent on the project.

INDEX

In this index, the entries have been placed in alphabetical order according to the letters in their natural sequence from left to right. Spaces, periods, hyphens, apostrophes, italicization, commas, and capitalization were ignored when establishing the order of placement of the index entry. Abbreviations and acronyms were not treated as though they were spelled out. According to a personal preference, indented sub-groupings were not used.

– D –

– I –

– J –

– K –

– L –

– M –

– Q –

– R –

– S –

www.ingramcontent.com/pod-product-compliance
Lightning Source LLC
LaVergne TN
LVHW090310210926
848480LV00001B/193
* 9 7 8 0 9 4 2 8 4 9 2 9 5 *